T0300648

APES ON THE EDGE

Animal Lives

Jane C. Desmond, Series Editor; Barbara J. King, Associate Editor
for Science; Kim Marra, Associate Editor

BOOKS IN THE SERIES

*Displaying Death and Animating Life:
Human-Animal Relations in Art, Science,
and Everyday Life*
by Jane C. Desmond

*Voracious Science and Vulnerable Animals:
A Primate Scientist's Ethical Journey*
by John P. Gluck

*The Great Cat and Dog Massacre: The Real
Story of World War Two's Unknown Tragedy*
by Hilda Kean

*Animal Intimacies: Interspecies Relatedness
in India's Central Himalayas*
by Radhika Govindrajan

*Minor Creatures: Persons, Animals, and
the Victorian Novel*
by Ivan Kreilkamp

*Equestrian Cultures: Horses, Human
Society, and the Discourse of Modernity*
by Kristen Guest and Monica Mattfeld

*Precarious Partners: Horses and Their
Humans in Nineteenth-Century France*
by Kari Weil

*Crooked Cats: Beastly Encounters in the
Anthropocene*
by Nayanika Mathur

*Dogopolis: How Dogs and Humans Made
Modern New York, London, and Paris*
by Chris Pearson

*The Three Ethologies: A Positive Vision for
Rebuilding Human-Animal Relationships*
by Matthew Calarco

Apes on the Edge

CHIMPANZEE LIFE ON THE
WEST AFRICAN SAVANNA

Jill Pruetz

The University of Chicago Press   CHICAGO AND LONDON

The University of Chicago Press, Chicago 60637
The University of Chicago Press, Ltd., London
© 2025 by The University of Chicago
Published 2025
Printed in the United States of America

34 33 32 31 30 29 28 27 26 25   1 2 3 4 5

ISBN-13: 978-0-226-83752-9 (cloth)
ISBN-13: 978-0-226-83751-2 (paper)
ISBN-13: 978-0-226-83753-6 (e-book)
DOI: https://doi.org/10.7208/chicago/9780226837536.001.0001

Library of Congress Cataloging-in-Publication Data

Names: Pruetz, J. D. (Jill D.), author.
Title: Apes on the edge : chimpanzee life on the west African
    savanna / Jill Pruetz.
Other titles: Animal lives (University of Chicago. Press)
Description: Chicago : The University of Chicago Press, 2025. |
    Series: Animal lives | Includes bibliographical references and
    index.
Identifiers: LCCN 2024019035 | ISBN 9780226837529 (cloth) |
    ISBN 9780226837512 (paperback) | ISBN 9780226837536 (ebook)
Subjects: LCSH: Chimpanzees—Behavior—Africa, West. | Human-
    animal relationships—Africa, West.
Classification: LCC QL737.P94 P78 2025 | DDC 599.885150966—
    dc23/eng/20240507
LC record available at https://lccn.loc.gov/2024019035

IN MEMORY OF MBOULY CAMARA

Contents

1 · Fongoli *1*

2 · The Fongoli Chimpanzee Community *23*

3 · Coping with Savanna Heat *43*

4 · Female *Pan* the Hunter *59*

5 · Risks on the Savanna: Snakes, Bees,
and Hippos, Oh My! *81*

6 · Neighbor Apes: Chimpanzees
in a Human Landscape *97*

7 · Conservation Threats and the Future
of the Fongoli Chimpanzees *117*

Acknowledgments *131*

Further Reading *135*

Index *139*

Color illustrations follow page 88.

· 1 ·
Fongoli

One of the most interesting behaviors I've witnessed at Fongoli involved three different chimpanzees attempting to hunt one bush baby.

Fongoli is a savanna site where my research team and I have studied wild chimpanzees since 2001. Our project was the first of its kind to habituate chimpanzees to the presence of human observers in a savanna landscape, so that we can follow these apes from dawn to dusk and collect detailed data on their behavior, as has been done with forest-dwelling chimpanzees for more than half a century. We have discovered a number of behaviors that either are unique to Fongoli chimpanzees or are at least rare when compared to chimps studied elsewhere. But hunting with tools is perhaps the most surprising of these discoveries.

In the instance I'm talking about, an adult female we call Lucille made a typical tool. She broke off a live tree branch about a meter long, trimmed off the side branches and leaves, and sharpened the tip with her teeth by whittling off pieces. She jabbed this tool repeatedly and vigorously into a cavity in a dead branch. She spent several minutes modifying her tool; given the way she worked at jabbing it into this cavity over and over again, it seemed obvious there was something in there. She finally abandoned the site and an adult male, Karamoko, who had been watching her, retrieved her tool and gave it a try. No luck for him either, and he abandoned the hunt. Finally, an adolescent male, Frito, tried his hand at tool-assisted hunting. The cavity was of a size that he could fit his smaller hand into it. However, each time he started to do this, he would jerk his hand back—again, as if something were inside.

Figure 1.1. Adult female Eva using a tool to hunt while daughter Tessa looks on. Photo by McKensey Miller.

When I climbed to look into this cavity, I saw what seemed to be a very angry bush baby!

Bush baby is the common name for a *Galago senegalensis*, a strepsirrhine or prosimian primate that is more closely related to lemurs than to monkeys or apes. Bush babies, like the one just attacked, are nocturnal and spend their days resting in hollow trees and branches. I could see a sort of shelf down in the cavity where he was partially covered, but he did have lacerations on his head from the tools. He had his mouth open in a threat and seemed to be pretty feisty, even with these wounds. I don't think the wounds were fatal ones, so I believe he survived another day, despite having three different chimps try to make a meal out of him.

This hunting behavior is unique to Senegal chimpanzees. While there are a handful of observations of chimpanzees using stick tools to hunt (a squirrel and a hyrax) at Mahale Mountains National Park, Tanzania, only at Fongoli is the behavior so frequent and systematic. At the time of this writing, we have almost six hundred observations of such spear hunting. Like most of the other behaviors that are unique to Fongoli chimpanzees (or are only rarely seen in apes elsewhere), the ultimate explanation for tool-assisted or spear hunting can be linked to the hot and dry environment in Senegal. Southeastern Senegal in May is brutal:

the rainy season is looming, so humidity is high, but the triple-digit temperatures that characterize the dry season have yet to subside. The result is a daily heat index upward of 49°C (120°F), and the nights are suffocating. Still, it's my favorite time of year because it's when the chimpanzees start their characteristic style of hunting.

Hunting is just one part of Fongoli chimpanzee behavior linked to the savanna landscape in southeastern Senegal. This is not just a book about how Fongoli chimps hunt. In addition to exhibiting unique or rare behaviors that stem from the challenges of living in an extremely hot, dry, and open environment, these apes are on the brink of extinction. At the time of this book's publication, scientists have classified the West African chimpanzee subspecies as critically endangered, and the Fongoli group is on the cusp of losing their habitat in southeastern Senegal. Even though the forces working against the chimps' survival in this area are linked to a human influx related to gold mining, local people are also negatively influenced by many of the same factors that threaten chimpanzees, such as the use of mercury and the increased potential for disease transmission because of the high human population in the Fongoli area now. Anthropogenic disturbances do not appear to be waning, and additional problems caused by climate change are currently impacting horticulture in the area to an even greater degree than such change is impacting the chimpanzees' behavior. Both humans and chimpanzees are struggling to deal with challenges in this area of southeastern Senegal, and as climate change continues to bring about unpredictable weather patterns, the livelihoods of each are in jeopardy.

My goal in writing this book is to highlight the unique or rare behaviors seen at Fongoli—especially the ones that enable these apes to exist at the literal edge of the chimpanzee range in Africa—and to highlight the precarious position these chimps are in because of their home here. Not only are chimpanzees in Senegal literally "on the edge" in terms of geographical range and environmental habitat, which ties in to many of their most interesting behaviors; they are on the edge of extinction as well.

WHY STUDY CHIMPANZEES AT FONGOLI?

The Fongoli chimpanzees are the first group of apes living in a savanna landscape that has allowed human researchers to collect detailed behavioral data. I began the Fongoli Savanna Chimpanzee Project in 2001 after surveying for chimps in Senegal in 2000. By 2005, We were successful in "habituating" the chimpanzee community: getting the chimps used to humans to the point, ideally, that they ignore our presence—a feat many thought impossible, because of the lack of success of previous efforts. For example, the Mount Assirik, Senegal, study in the Niokolo-Koba National Park in the 1970s was abandoned after four years because of a lack of progress. I now have nineteen years of behavioral data on this community of chimpanzees, and I have recorded a number of unique or rare behaviors. And, although the Fongoli apes also exist in a human landscape, sharing their space with people outside national park boundaries, they are challenged in many ways by their close proximity to people. Yet humans are part of the chimpanzees' lives, and we all have likely been coexisting for millennia. Most long-term chimpanzee studies are in protected areas where they are isolated from humans—not necessarily a natural situation in our human-modified world. (I'll talk more about this subject in chapter 6.)

Although the inclination might be to think that chimpanzees living alongside humans face many disturbances, some aspects of the Fongoli chimpanzees' lives are less negatively influenced by humans compared to protected study sites. For example, in Kibale National Park, Uganda, a number of chimpanzees in the Kanyawara community show deformities resulting from being caught in snares. Wire snares entrap the hand or foot of a chimpanzee, even though the snares are being set to trap small antelopes. While chimps may be able to extract the wire from the tree it's attached to, they can rarely extract the snare from their limb until it has caused some of the tissue to rot off. Researchers and wildlife service authorities will try to intervene and remove the snare, but the chimps are usually left with digits or parts of their limbs that ultimately shrivel and fall off as the blood supply is cut off. Many chimps are able to survive with such deformities; they are of course limited in some activities. Fortunately, at Fongoli, we have recorded only three snares since 2001, and head researcher Michel Sadiakho was the only one who ever stepped in one (he was basically unhurt, though bruised).

Similarly, living alongside humans suggests that the apes here might not be subject to predation pressure experienced in nationally protected parks. However, although the predator diversity and density at Fongoli is relatively low compared to the Niokolo-Koba National Park in Senegal, where unhabituated chimpanzees live, the leopards (*Panthera pardus*)

and spotted hyenas (*Crocuta crocuta*) make Fongoli relatively more predator rich than some national park sites where chimpanzees have been
studied for decades, such as in Kibale National Park in Uganda.

The savanna biome at Fongoli is definitely distinct compared to the
forests where almost all other long-term studies of chimpanzees have
taken place. In 2018, the Issa chimpanzee community in Ugalla, Tanzania, a Miombo woodland savanna environment, was habituated to observer presence, making it only the second site where apes in a savanna
landscape can be followed daily by observers to collect detailed behavioral observations. Chimps living in forests, however, have been studied for sixty years at some sites, and there are more than a dozen long-
term study sites of forest-living chimpanzees. About half of these can be
found in Uganda alone. Although chimpanzees may live in lower densities outside of forested areas, they have also likely been using savanna
landscapes for as long as they have been chimps.

Many people think of an open grassland when they hear the term *savanna*, but that is only one type of a savanna. Woodlands dominate the
savanna landscape at Fongoli, and this is where most of the chimpanzees' foods, such as baobabs, shea trees, and other fruiting trees, are
found. Grassland is the second most common savanna type; you'll also
find some trees here, but they are usually small and shrubby. Grasslands
don't offer a lot to chimpanzees in terms of foods. Other parts of the Fongoli savanna mosaic include bamboo woodland, human features (trails,
villages, farm fields, dirt roads, small gold mines) and gallery forest.
Gallery forest accounts for only about 3 percent of the apes' home range
area of more than 100 square kilometers (about 70 square miles), but it
is a crucial type of habitat—this is where you find water sources that last
throughout the year.

We have names for each of these forested areas. One of these is
the seasonal Fatako stream, or marigot (a French word that we use to
describe these creekbeds, which are dry most of the year), less than
500 meters long but relatively deep and full of water for several months
of the year. It is the closest thing to the deep valleys that characterize
the Assirik area of Niokolo-Koba National Park. There is also a tiny
patch of gallery forest near the village of Djendji, which begins below
a plateau and extends down about a hundred meters to a small spring.
During the dry season, the spring is more aptly described as a puddle,
but it is there year-round. Other bits of gallery forest begin below the

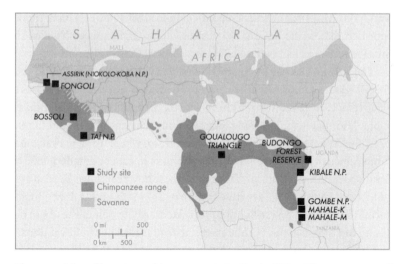

Figure 1.2. Map of long-term chimpanzee study sites in Africa. Figure courtesy of
Todd O. Pruetz.

edge of plateaus, where water flowing off these flat savanna grasslands
over the years has eroded out a ravine that is now full of forest. One of
these is the Sakoto ravine; another is Maragoundi. Parts of the Kerouani,
Tukantaba, Oundoundu, and other seasonal streams are the only other
significant bits of evergreen, somewhat closed gallery forest habitat in
the 40-square-kilometer or so core part of the study area that the chimps
use most over the course of a year.

A lot of the Fongoli area includes more open marigot or streambed
woodland habitat, so that over half of the home range of the chimps includes some type of habitat with trees. Part of the Fongoli range also
includes the Gambia River, but the chimps don't use this area as much
as you might expect, given that it is a permanent source of water during
the dry season. Perhaps they avoid it because, as people here describe it,
it has bad water—not potable—compared to other water sources, such
as natural springs.

Stresses like heat and other climate-related factors that characterize
the Fongoli chimpanzee environment is what makes these apes so interesting in my opinion—and so useful for informing us about theories
of our earliest bipedal relatives (hominins—the earliest bipedal or two-
legged apes in our lineage, living between five million to seven million
years ago) in a similar climate. Studying apes in this type of environment

Figure 1.3. Mbouly Camara (*right*), author Jill Pruetz (*center*), and Dondo "Johnny" Kante (*left*) during the early years of the project at Fongoli. Photo courtesy of the Fongoli Savanna Chimpanzee Project.

broadens our understanding of chimpanzee behavior and ecology in general and gives us insight as to how primates exist on the very limits of their range. This is especially important as climate change issues negatively influence the ecosystems of primates, including humans, worldwide; the edges of a species range is where a population usually lives in lower densities, having reached the limits of an adaptive niche.

It's popular perception that we know pretty much all there is to know about chimpanzees. I've often said in public lectures that chimpanzees are the best-studied wild mammal, and I haven't been challenged on that point yet. If you add up the various long-term research sites where these apes have been studied for at least a decade and take into account the length of the studies, we have accumulated more than five hundred years of observer effort studying chimpanzees. Yet, primatologists will tell you that if new study sites were to be established today, formerly unknown behaviors would most certainly come to light. One of the most recent chimpanzee communities to be habituated—located in the Loango National Park in Gabon, West Africa—was observed to engage in aggression toward Western lowland gorillas (*Gorilla gorilla gorilla*), killing an infant gorilla in each of two interactions. Chimpanzees at a site called Goualougo, in the Congo, is one of the few sites where central African chimps (*Pan troglodytes troglodytes*) have been habituated for long-term study, and they have been seen to use more complex tool sets than chimps at most other sites—including using a sequence of different tools to open stingless beehives and then extract honey, as well as using specific tools to open access tunnels in termite mounds, then switching to a different tool for termite fishing. We know but a part of the entire behavioral repertoire of chimpanzees, and, given the rapid decline in their numbers plus our caution about habituating new groups, it's unlikely we'll ever know the full complement of behaviors exhibited by our closest living relative. This makes it even more important to continue long-term studies on already established groups.

STARTING A LONG-TERM RESEARCH STUDY SITE

Despite now working with chimpanzees for more than thirty years, I originally didn't want to study them. *Everyone* seemed to want to study chimpanzees, and I am a sucker for the underdog, whether it be a struggling sports team or a less-studied scientific species.

What I didn't count on was coming face to face with chimpanzees and having them sidetrack my best-laid plans. In between graduating with a bachelor's degree in anthropology and one in sociology from Southwest Texas State University (now Texas State University) and graduate school at the University of Illinois Urbana–Champaign, I volunteered at a nearby primate facility. At that time, the American Society of Primatologists published a print newsletter that included job and volunteer opportunities. I was surprised to learn that there was a National Institutes of Health breeding facility for chimpanzees less than an hour away from my university. I began as a volunteer behavioral psychologist and was hired into the same position after a few months. I worked there for a year before leaving for Peru, South America, to start my graduate career.

That year with the chimps changed my life forever. I believe it was one of the happiest years of my life. My job was to encourage species-specific behaviors, such as being more active, and reducing stereotypical behaviors and inactivity in the captive chimpanzees—generally, to enrich their lives. In one study, I measured how effective different types of objects were in getting the chimps to play. One of my fondest memories was of two older females taking away enrichment items from the youngsters to play with themselves. It was heartbreaking for me to leave, and my work on moustached tamarin monkeys (*Saguinus mystax*) in Peru never did measure up after that. Once I met a chimpanzee face to face, it was all over. It was only a matter of time before I got back to these apes one way or another. I wanted to study wild chimpanzees, and I wanted to start my own research site. It took me only a decade to get there.

Most students that get into the field of primatology see only a small part of the story of how one goes about beginning research, which is unfortunate. I recommend Craig Stanford's book *Chimpanzee and Red Colobus* to students because not only is it a good book on the predator-prey dynamic, but it also describes aspects of the scientific process that are less accessible to students. I think students become disheartened by the lack of information on how to achieve what is a difficult objective even for experienced scientists: convincing someone that you, a novice, are qualified to go out and conduct original research that none of the older, more experienced and superbly intelligent scientists have thought to do. You want someone to give you money to do this research even though you have in no way proved yourself at all? However, if you write an amazing proposal, you will win said sum. In fact, even for most senior scientists, the road to finally embarking on a research project is fraught with difficulties that never really go away—such is the nature of fieldwork.

But even though the road itself may be convoluted (at least, mine was), I think it's useful for students or other aspiring researchers or conservationists to know these stories. I believe that passion and perseverance were key to my achievements in and outside academia, acknowledging that my white, middle-class upbringing in the United States made it easier for me. I've written of my circuitous arrival in primatology elsewhere, but, to summarize, I took an anthropology class with my roommate in college and was hooked; I changed my major and got into archaeology first. I loved fieldwork, but not cleaning potsherds. So I ventured further, into primatology, which had also resonated with me. I volunteered to work with chimpanzees, a species I did not intend to study purely because of their popularity among students of primate behavior.

This was when I followed my theoretical interests of studying tamarin monkey social behavior—but the chimps had ruined me. I fell in love with the species and there was no turning back. I took a relatively quick route to a dissertation project in managing a study on patas monkeys and vervets in Kenya, where I also conducted my doctoral research. While I thoroughly enjoyed my savanna-monkey-study subjects, I was still determined to begin my own study of chimpanzees. I did love the savanna where I lived for two years in Kenya, so decided to combine that love with my dedication to the study of chimpanzees. But no one had yet been successful in habituating chimps in a savanna so that they could be studied systematically as Jane Goodall and now many others have done with forest chimps. No matter—I knew I could.

As a postdoctoral research scholar, I traveled to Senegal with primatologists Dr. Linda Marchant and Dr. Bill McGrew to conduct a survey of chimps in Senegal. They focused on collecting material for a genetic study, though we all worked together to complete our respective projects. Bill was one of the few people who had studied chimps living in a savanna: as part of the Stirling African Primate Project, Bill and others embarked on a four-year study of chimps and other primates at the Assirik site in Senegal, within the Niokolo-Koba National Park. They abandoned the project in the late 1970s when it seemed clear the chimps were not becoming habituated—at least, not according to a trajectory based on forest-dwelling chimpanzee research at Gombe Stream Research Centre, Tanzania, that Bill was familiar with. During my postdoctoral research, I also surveyed chimpanzees outside the Niokolo-Koba National Park, where the Assirik site was located, and it was here I saw the difference between the behavior of apes within the national park and those living outside it, alongside people who were predominantly farmers and pastoralists but who also did some hunting and gathering. Chimps within the park rarely encountered people and seemed terrified of us; chimps outside the park were also afraid, but less so. They were used to people as part of their environment, even though humans were treated as a threat (and to some degree a predator, as I would come to find out in my studies and those of students working at Fongoli). I realized my best bet was to attempt habituation at a site outside the park. As an anthropologist, I was also intrigued by the fact that chimpanzees coexisted with people in Senegal with a large degree of success. A master's student, Kerri Clavette, would later research the cultural taboos we were told prohibited the systematic hunting and eating of chimpanzees in Senegal. Such hunting elsewhere is problematic for the conservation of this species at some sites in Africa, especially where chimpanzees raid crops.

As a postdoc at Miami University in Ohio, I began to apply for grants to fund a long-term research project in Senegal at one of the most promising sites I had surveyed in 2000, not far from the town of Kédougou and along the banks of a seasonal stream called the Fangoly (I anglicized the word when I named the chimpanzee study community after it to reflect what the local word would sound like in English). Getting funding to start a long-term study was another issue. After writing about nine proposals, I earned funding from two organizations—the National Geographic Society and Primate Conservation, Inc.—and started a tenure-track position at Iowa State University in 2001. The grant reviewers noted the difficulties I would have achieving my goals, skeptical that I could do what others had failed to do—namely, habituate chimpanzees in this savanna environment.

However, I was confident I could succeed. One of my strengths (and perhaps a weakness in some contexts) is not giving up. Luckily, I had the support of my university, Iowa State. I was also encouraged to submit a grant to the National Science Foundation (NSF) under its "high-risk" category, which would recognize that the proposed research might not result in the desired outcome. The high-risk NSF grant allowed me to try to habituate these chimps, and subsequently I was able to continue earning grants to study aspects of chimpanzee behavior and ecology in a savanna environment and ultimately to habituate the Fongoli chimps. I later received funds from the Leakey Foundation, the Wenner-Gren Foundation for Anthropological Research, the American Society of Primatologists, the Great Ape Trust (now the Ape Initiative), and the US Fish and Wildlife Service. The National Geographic Society has awarded me more grants than any other funding organization, and they also named me an Emerging Explorer in 2008, which entailed a monetary award to support my research.

FINDING FONGOLI

Senegal is the westernmost country in Africa. It lies below the Sahara Desert, with the northern part of the country including part of the Sahel, or the semi-desert bordering the Sahara. The Senegal River, which forms the national boundary between Senegal and Mauritania, to the north, is also roughly the boundary between the Sahara and the Sahel.

Senegal was colonized by France in the early 1900s but gained independence in 1960. Although much of the wildlife was decimated after colonization, Senegal also has a history of attention to conservation, with its first national park established in 1925. The country has now experienced drought for over half a century, affecting humans and wildlife alike. Chimps in Senegal are found only in the southeastern part of the country; their range extends into Mali in the east and Guinea to the south. The environment here is called the Sudano–Guinean zone and includes typical savanna but also the crucial patches of forest where chimps' water sources are found.

It made complete sense that I would combine my love of chimpanzees with the hot, dry environment of southeastern Senegal, and I kept repeating that to myself as we drove into this unbelievably hot and barren environment where it was hard to fathom that furry chimpanzees would dare to live. If it wasn't for the fact that we saw old chimp nests while driving on the dusty gravel track to Mount Assirik, I wouldn't have believed that chimps could survive in southeastern Senegal.

We arrived in February 2000, just as the dry season really gets going. Almost all tree species have lost their leaves, with shade being confined to the handful of evergreen plant species (*Cola cordifolia*, *Saba senegalensis*) that are found in gallery forest and ecotone at the edges of plateaus where the water runoff has produced tiny patches of forest. The dead leaves characterizing the remnants of chimpanzee nests are easily seen at this time of year, and even the woven "bowl" (which is the only remnant branch framework of long-gone nests) can at times be seen for years after being used. Temperatures are in the triple digits for months, but it's a dry heat. The heat index usually doesn't top 49°C (120°F) until May, when the rainy season threatens and humidity increases.

During my survey in the year 2000 of the Niokolo-Koba National Park and outlying areas in southeastern Senegal, I found two sites outside the park where chimpanzee-nest densities indicated a relatively healthy population of chimpanzees, similar to what we found in the national park. When I returned in 2001 to initiate my long-term study, I kept several options in mind, including what became the Fongoli site. I decided that Fongoli was the most promising site, after working with a Malinke village farmer and hunter, Mbouly Camara, who had helped me with the previous year's survey at a seasonal village most people called Giringoto (pronounced "gidingoto" in English), which roughly translates to "the place where you can cross the stream at the fallen tree." Mbouly later asserted that the formerly seasonal village was also now called Fangoly, and he loved the area so much that he asked to be buried outside this small village, rather than at his larger, permanent home village or the site of the former village at Maragoundi, where he had been born, but which had been abandoned because of bad water. I also hired a guide, Dondo "Johnny" Kante, from a neighboring Beudick village, Thiobo (pronounced "chobo" in English), to translate my French into Malinke. (Everyone in Senegal speaks more than one language, and most people speak several. Dondo speaks nine languages and learned English after starting work at Fongoli. He once lamented that he didn't think he could learn a tenth!)

Another site with high nest density ended up being significantly deforested in following years, and it was further south than Fongoli, near the border with Guinea. At this and a few other sites I kept my eye on, I worked with a guide who lived in Kédougou by the name of Jean Pierre Camara. Jean Pierre was also an excellent guide and knew so much about the natural history of the area. Even though I ultimately chose the Fongoli site, I kept in touch with Jean Pierre, and my nonprofit organization Neighbor Ape helped his oldest daughter complete school and then get a degree in pharmacy studies in Dakar. Dondo continued working for me and ended up becoming my project manager. I also liked the fact that Fongoli was further north in Senegal and had less forested habitat and more open woodland and grassland.

As an anthropologist, I was cognizant of the importance of working with local stakeholders, as they are termed by development agencies. It would be impossible to conduct a successful research project in Senegal if I didn't have the support and assistance of the people who live there. Conducting fieldwork in another country or area or culture is sometimes challenging, to say the least. But one of the most exciting things about it is those differences you encounter. The expertise of local people is invaluable, and, besides seeking their assistance for information on the ecology of the area and what they knew about the chimps, I needed their informal permission to work there. Plus, collaboration with authorities at different levels was instrumental in continuing to protect the chimpanzees and their habitat, especially since rapid anthropogenic changes characterize Senegal, as they do most areas of the world.

With the blessing of Mbouly, who acted as village chief and became my senior field assistant/researcher, I continued my study of the Fongoli chimpanzees, and we eventually habituated the ape community to being observed by researchers. We established camp at the village of Fangoly, which had been a temporary farming village during the rainy season but became a small, permanent settlement when we set up shop there—Mbouly and his extended family stopped returning to their larger, nearby home village during non-planting and non-harvesting seasons. After three failed attempts (due to cave-ins) we were able to successfully install a well for clean drinking water, and the Fongoli Savanna Chimpanzee Project has been based there ever since. Now, it also serves as the home base for the Bantankiline Chimpanzee Project, directed by Dr. Papa Ibnou Ndiaye, a professor of biology at Université Cheikh Anta Diop in Dakar, Senegal. The Bantan chimpanzee community is adjacent to the Fongoli chimp community; while we have studied them indirectly over the years, they are not habituated to human presence, and we do not have plans to do so. There is much to be gained from indirect study, and habituating them would mean we would need a permanent and continuous presence at Bantankiline as well as at Fongoli, in order to provide for the chimpanzees' safety. Even though our Fongoli team is small and the number of "strange" humans allowed out with the chimps at any one time is few, the chimps may at some point in the future become generally habituated to humans, which could put them in harm's way.

I have always tried to maintain a small footprint at Fongoli, for a number of reasons. In part, this is to try and minimize our effect on the ecosystem at Fongoli. Since we live in a village and depend on the people of this area for the success of our project, I have purposefully tried to live as they do, with relatively simple housing and facilities. We have added some amenities, though. In addition to a well, we put in a long-drop toilet, which consists of a deep hole in the ground and a bamboo screen for privacy. Each of our six huts (one kitchen–library hut and five bedroom huts) have cement floors, while some villagers have only dirt floors. We eat our meals after returning from the field because I pay a family to make extra food and put aside a portion of for us when they cook their own meals. We have also thus far succeeded in keeping the chimpanzees habituated to just our small team of observers, so they are not as susceptible to disease transmission and potential poachers, as is the case where apes become habituated to humans in general. Fortunately, I have been part of a team of very talented researchers who have worked for the Fongoli Project for years.

FONGOLI CHIMPANZEES

As I mentioned, one of the reasons it was difficult to secure funding for the initial year of research was due to the perception that chimpanzees in a savanna landscape could not be habituated to human presence. And yet we were able to succeed.

A saving grace at Fongoli was the apes' reliance on only a handful of permanent water sources available to them during the dry season. We began the habituation process in 2001 by conducting water source vigils at one particular water point (we call it Point D'eau) near the largest village within the Fongoli chimps' home range, Djendji. This seep becomes a small stream during the rainy season, but it shrinks to a puddle during the peak of the dry season.

Our method was to spend five to seven hours sitting some distance from the seep—which is hidden by a set of large boulders at the bottom of a steep slope—so that the chimps could see us as they arrived to drink and we could begin to observe them. If we saw individuals descending the slope to approach the water source but seeming hesitant to continue their approach and drink with us sitting nearby, my protocol entailed leaving the vicinity of the water point for an hour and then returning, giving these shier individuals time at the water source alone.

We made progress more quickly than I thought would be the case. Even during the first several months we could identify one or two individuals in the community. By the third year, we could identify fifteen different chimpanzees, which included most of the adults in the group. (It wasn't until early 2006 that I was able to identify all the adult females in the community and their offspring, though.)

Getting the chimps to let us follow them away from the water source was a different matter entirely. It wasn't until after four years of habituation effort, in early 2005, that we were able to follow adult males from night nest to night nest—in other words, from when they emerged from one nest in the morning to when they built another nest that night. Chimpanzees, like all great apes, sleep in a nest made of broken or bent-over branches that are then intertwined and sometimes lined with more branches or leaves to make a leafy platform. Usually they make a new nest each night, although chimpanzees at Fongoli do reuse nests quite often—sometimes their own but others' old nests too.

There are varying reasons to limit habituation of a group of chimps.

When I began my research at Fongoli, I was interested in studying the behavioral ecology of females in this relatively harsh (for apes, anyway) environment—that is, how the environment shapes and influences an individual's behavior. Early on in the establishment of the Fongoli Savanna Chimpanzee Project, I learned that female chimps were at the risk of being poached by hunters because of the slight but real chance that humans would target them in order to try and acquire their infants for the pet trade. We actually had one such case in 2009, but we were able to confiscate the baby and successfully return her to her mother within five days. At the beginning of the study, I decided I wouldn't habituate female chimps at Fongoli to being followed by us. Even though females are calm in the presence of human observers when they are in large parties or subgroups—when male chimps are around—they are timid if humans encounter them on their own. Accordingly, I now collect data on females only when they are in mixed-sex social groups, and I don't allow them to be singled out for research, all in an effort to prevent overhabituation.

In other words, we follow a single male (if possible) from the time he gets up in the morning to when he goes to sleep at night, taking data throughout the day—he is the focal subject. But we don't follow females this same way. Males are now very well habituated to us—allowing us to follow them as if we weren't even there most of the time. Females are also habituated to us when they are in the same social subgroups, or "party," as males, which is most of the time at Fongoli, since these chimps are more cohesive than chimp groups at many other sites. So, we can collect data on female chimpanzees as well, but they are not followed around doggedly as are the less vulnerable adult males. Regardless of who the chimp is, we always try and maintain a 10-meter distance from them as well. This cuts down on the likelihood that we can spread germs to them, and it helps us influence their behavior as little as possible.

Chimpanzees are well aware when they are being followed, even though they may not give much sign of it once they are habituated. And, even though females are relaxed in mixed-male-and-female subgroups, when we do encounter females on their own, with their offspring, even after fifteen years of observational study, they are much more nervous—despite the fact they are incredibly calm when we see them in big subgroups. In that case, we just record the females we see and where they are, and then we move on to look for males. I hope that this keeps females more vigilant toward humans, and so far it seems to be the case.

Since we can never predict what will happen, we also try to ensure that all the Fongoli chimpanzees are somewhat wary of humans in general. This is why I have a relatively small research team. I also have only a few students per year conduct research on the Fongoli chimps, and they must accompany one of the full-time Senegalese research assistants, who the chimpanzees know very well. Since the chimps were reported to hunt with tools in 2007, I have had numerous documentary film teams seeking to record this and other behaviors unique to Fongoli or rarely seen at other sites. Usually, I allow only one film crew on site in any given year, and they also have to be with a research assistant at all times. We have rules about how long chimpanzees can be filmed in a given time period and how many people can be out with the chimps at any one time. All these rules are in place to keep the chimps safe in an area that is not officially protected in Senegal. While the chimps are officially protected in this country by laws (and via cultural taboos against eating them), recent changes associated with the ongoing gold rush in Senegal means that people from all over now live in the area (and they may not have the same taboos against eating apes that people in Senegal have).

Even in May 2001, about a month after I started the Fongoli Project, we were able to recognize some individuals at the water source. The very first chimp we gave a name to was Ross. I wouldn't be surprised if he had been more than fifty years old when we first knew him (and we now know that chimps can live to this age in the wild as well as in captivity). He couldn't see or hear well, and he had only two teeth left in his mouth, both in his bottom jaw. Even local people knew of him; I assume they often saw him, while he probably didn't see them. His face—very splotchy, with pinkish-white areas of depigmentation that indicate age— even graced the pages of *National Geographic* in April 2008. He likely habituated to us because he was drawn to the water source and resting nearby; he seemed to worry little about our presence. I have to say that even though Ross disappeared in 2009, for a long time I just told local people who asked about him that he was doing fine because I didn't want to be the bearer of bad news. Ross was named after the friend of a friend of one of the students studying at Fongoli at the time, someone who appeared prominently in stories told around mealtime. Over the years, I have asked students and other researchers to provide names for chimps, with the contingency that they not be named after anyone too famous, to avoid the danger of ascribing personality traits of that person to the chimp. Now I ask donors to my nonprofit organization to provide names, and the Senegalese researchers take turns choosing from this list.

I recorded seeing chimps six times at Point D'eau in May 2001, and one adult female in particular appeared relatively calm and curious. This turned out to be Farafa, who had her infant son, Frito, and her juvenile son, David, with her most of the time. She also had an adolescent female accompanying her, but we were unable to get a positive identification on this young female. In order to positively identify an individual chimp, we want to see all parts of their body and to describe them according to specific traits, like scars or natural markings, in addition to being able to estimate their age and identifying their sex. We were not able to do this with that young female, and she likely left the group soon after our study started, since she was an adolescent and it is typical for female chimpanzees to transfer into other social groups at maturity, at Fongoli and elsewhere. About 25 percent of females at Fongoli do stay within their natal group; I assume the others move on to the adjacent and unhabituated Bantan chimpanzee group to the north and perhaps to other chimp communities from there.

Sometimes a young adult male was with Farafa's group as well—this turned out to be her older son, Mamadou, confirmed via mitochondrial DNA analyses conducted by Fiona Stewart at Cambridge University. Mitochondrial DNA (or mtDNA) stems from the mother's line only, so this allows us to identify relatives through a mother's line but not the father's. We gave him the name Mamadou. Based on Mamadou's physical appearance, I estimated that he was a young adult, probably around sixteen or seventeen years old, at the beginning of our study. If he was Farafa's firstborn, using estimates of ages at which females first give birth at other sites, which is around thirteen years, Farafa had probably been born in the early 1970s, and had given birth for the first time around 1990 (at the time of publication, in 2024, she would be in her fifties).

Mamadou holds a special place in the annals of Fongoli history, in part because of his attempts to discourage us from following him during the habituation process. Early on, he rolled a boulder down toward one field assistant. Another time, he woke up "on the wrong side of the nest" and gave an aggressive intimidation display toward a researcher first thing in the morning—surprising him so that the observer made the mistake of stepping back—a clear sign to Mamadou that he had the upper hand, and so he kept coming. Displays are typically feats of running or otherwise charging around, with hair on end, throwing whatever is at hand, in an effort to intimidate someone. Mamadou got to within a meter of the assistant when a standoff took place, and Mamadou ultimately went on about his business. There were a few other memorable times Mamadou attempted to put humans in their place. Even after we'd been studying the Fongoli chimpanzees for years, he would sometimes try to lose us during daylong follows—and sometimes he did. It all made him ever more endearing as the crotchety tough guy he was. He was the beta male (second in rank in the group) most of the time we knew him, except for a brief yet significant fall to the bottom of the male-dominance hierarchy: after a coup that ousted the alpha male at the time, Foudouko, Mamadou appeared to either break or dislocate his leg at the hip during one of the battles that involved all the adult males.

The Fongoli chimp group is characterized by more males than females relative to any other habituated chimpanzee social group. This might explain some of the behaviors we see that are uncommon or absent elsewhere, and I focus on some of these in later chapters. The community itself is on the small size for chimpanzees, averaging thirty-two individuals over the course of the past fifteen years, and we have always had between ten and twelve adult males in the social group at any one time. We have thousands of hours of data on these adult males now; like most primates and other animals, in fact, they vary widely as to their personality.

RESEARCH AT FONGOLI

Following habituation, we began to discover behavior after behavior that was unexpected, and this has been what has guided my research. The Fongoli Savanna Chimpanzee Project is, on one level, a study of the natural history of apes in a challenging environment. There are many differences between chimpanzees here compared to those living in more forested habitats, and new avenues of study continue to appear. Even though I approach my research with the hypothesis-testing paradigm—characteristic of modern scientific studies of wild animals—I often find myself shifting my focus to other areas because of new discoveries of chimpanzee behavior. Even events like the chimp-napping by poachers of an infant, Aimee, and our subsequent rescue and return of Aimee to her mother (which I talk about in chapter 6) has led me to new areas of study, including chimps' capacity for empathy.

I am continually surprised that as I discover something about their behavior that is unexpected (at least, according to the literature on chimpanzee behavior), the Fongoli chimps almost always lead me back to something to do with this harsh environment. They use caves and water to cool off in their extreme environment, unlike apes at other sites (I describe this in chapter 3). This makes all the heat, sweat, heat, insects, heat, illness, more heat, et cetera, worth putting up with. I already mentioned that the transition from the dry season to the rainy season is actually my favorite time of year in Senegal, mainly because this is when Fongoli chimpanzees start hunting bush babies. Even though the heat index makes us all pretty miserable and the malarial mosquitos start to come out in force, the leaves on almost all the tree species are starting to flush, so that little by little, shade appears again, and grasses on the savannas emerge as a green carpet. Usually there is a big rain, followed by a week or two of stifling-hot temperatures. But the relief of the rainy season is in sight. In April, at the end of the dry season, you can easily imagine the world burning up in some apocalyptic scenario. You imagine yourself, the chimps, and all the rest of life in sight helpless in the face of the spontaneous combustion of southeastern Senegal. Then clouds start to appear and the heat becomes even more unbearable, but at least there is hope. The chimpanzees pant-hoot with excitement to hear the first sounds of thunder each year as the dry season fades. One year, I literally heard the alpha male, David, start a chorus of excited screams—rather than just pant-hoots—as distant thunder boomed. I have had to restrain myself from pant-hooting along with them at these times more than once. I imagine my thoughts are a little more dramatic than the chimps: "We aren't going to die! It really is going to rain, and we will all *live!*"

At Fongoli, chimpanzees live alongside more humans but fewer other mammals than, for example, apes in the Niokolo-Koba National Park, which is just about 30 miles away. However, while many of the species of large fauna at Fongoli have been extirpated, chimpanzees here still must face predation pressure and competition with other species. For example, at the Assirik site in the Niokolo-Koba National Park, the baboon (*Papio papio*) density is much higher than at Fongoli and other areas outside of the national park, so apes there theoretically face higher competition over different foods. Around Fongoli, monkey species such as baboons, patas monkeys (*Erythrocebus patas*), and green monkeys (*Chlorocebus sabaeus*) are considered crop pests and are hunted by some people. This means that the prey availability of these species is lower at Fongoli than at the Niokolo-Koba National Park, since each of these primates is included in the Fongoli chimp diet. The high degree of anthropogenic pressure at Fongoli adds other levels of complexity to their ecosystem that is largely lacking in the national park, where humans were removed in the 1960s. (I talk more about the relationships chimps have with the local people in chapter 7.)

Even though lions (*Panthera leo*) and African wild dogs (*Lycaon pictus*) have been exterminated around Fongoli, the apes here must nonetheless face some predator pressure. We first became aware of a resident female leopard in 2002, when my project manager, Clayton Clement, along with Mbouly, found a bushbuck carcass cached in a cave at Sakoto ravine, indicative of a leopard stash. In 2012, I had set up a camera trap at the Sakoto pool, where chimps like to soak in the early rainy season. I removed the camera card as I was leaving Senegal and didn't get a chance to review the images until I was teaching at a primatology field school in a different country. I was shocked to see multiple images of an adult female leopard passing the camera trap several days in a row, moving away from the cave area at dusk and back toward it at dawn. She was only a few meters from our major path to and from Fangoly village! Another time, one of my field assistants, Waly Camara, observed an encounter between the chimps and a leopard at Djendji ravine, where there are also small caves (see chapter 5). We have also observed the Fongoli chimps throwing stones at spotted hyenas as well as at baboons—a behavior we see even during fights within the chimpanzee social group. The pressures that the Fongoli chimpanzees face as far as predators and nonhuman competitors are concerned, however, are dwarfed by those associated with the increasing human population in southeastern Senegal.

· 2 ·
The Fongoli Chimpanzee Community

While the transition from the late dry season to the early rainy season is one of the toughest times for researchers to be at Fongoli—the high heat and high humidity routinely make for a heat index upward of 49°C (120°F)—it is also one of the best times to observe the chimps. The trees have started to leaf out, and, with the first rains, the grass begins to crop up on the open plateau, which has looked like the surface of the moon for most of the dry-season months, after fires sweep through. This makes for great visibility as well as great observation, because the entire Fongoli chimpanzee community ranges together as a single group for most of the early rainy season. This means we can follow thirty-plus chimpanzees in an often noisy, raucous group. Even adults begin to play together as the temperatures plummet with the wild thunderstorms that characterize this time of year—a storm can bring down temperatures by about 11°C (20°F), and it is exhilarating to feel!

One of my favorite sights at this time of year is a game I call "catch-foot" that the chimps play as they travel, often across the wide expanses of a plateau. The chimp behind grabs the foot of the individual walking in front of them and catches it as they are both moving; the one in front pulls their foot away and keeps moving, and the chimp behind continues to grab at and catch their feet as they travel. For some reason, it is especially enjoyable to see adult males play this game, and the adult males also play catchfoot with adult females and vice versa. I love seeing the oldest chimps, Farafa, Bandit, and Siberut, playing this way; you don't see this in the energy-draining dry season very much. The high degree of sociality at this time of year is always exciting, and we likely take these large social groups (or "parties," in primatology parlance) for granted at Fongoli, as this is not a scene that's observed at all chimpanzee study sites.

SOCIALITY ON THE SAVANNA

As social primates, humans may take for granted that most living animals are not social—at least in comparison to primates. Being social (rather than just what I consider being gregarious, like a herd of thousands of migrating wildebeest in the Serengeti) means you know your social group members and have some sort of relationship with them, whether it is more positive or more negative. Almost all primate species live in some type of social group, and chimpanzees are no exception. Chimpanzees at Fongoli differ in some respects from those at other sites regarding their sociality: Fongoli chimps range in large, cohesive groups much of the year, and their group has been consistently characterized by a larger number of mature males than females. At almost all other long-term chimpanzee study sites, adult males outnumber adult females, and the chimpanzee community—as some primatologists call chimpanzee social groups—rarely ranges together at one time as a cohesive unit.

Sociality shapes many aspects of a chimpanzee's life and, at Fongoli, almost any behavior that is studied, including the famous tool-assisted behavior by females, must be considered within a social context. Sociality has many levels, and understanding the nature of relationships between and within chimpanzee groups sets the stage for better interpreting the differences we see between Fongoli chimpanzees and chimps studied elsewhere.

Species that live in fission–fusion societies, like chimpanzees and bonobos (*Pan paniscus*, the sister species of chimpanzees, which live only in the Democratic Republic of the Congo), are routinely defined as exhibiting a loose community structure in which all individuals are rarely seen together at any one time. In chimpanzees, individuals know one another in their community and have some sort of social relationship, but the relationships between different communities of chimpanzees are generally hostile, especially between adult males of different communities. Chimpanzees at Fongoli appear to be somewhat of an exception to this pattern, although their social structure generally fits a fission–fusion model. While chimps at Fongoli exhibit most typical chimpanzee behaviors, there are some aspects of their social structure and behavior that do differ from chimps living in forested areas, and again, many of these differences can be traced back to the semiarid savanna environment here.

Party size or subgroup at Fongoli is relatively larger, on average, than at most other sites where chimps have been studied. During the dry season, when water is scarce and foods around water sources are then depleted first, chimpanzees have to range further from these sources to eat, and this is when we do see smaller subgroups or parties. During the late part of the dry season especially, scarce water sources effectively limit the otherwise relatively abundant food sources by restricting the chimps to a few core areas within the community's large home range (about 100 square kilometers, or 70 square miles). But, in general, chimpanzees at Fongoli—like West African chimpanzees in general as well as savanna-dwelling chimpanzees at the Issa, Tanzania site—range in relatively larger subgroups than chimpanzees at other sites; say, more than a quarter of all independent chimpanzees can be found together on average. The exception to this pattern is when an estrus female effectively reunites most of the community, which can happen at any time of the year. During the early rainy season, when water is relatively evenly distributed, the community regularly ranges together as a whole for four to eight weeks or more.

Explanations for the cohesiveness of the Fongoli community include (1) a skewed sex ratio (perhaps causing males to track the relatively few females more intensively than at other sites), (2) a difference between West African chimpanzees and other subspecies, (3) an abundance of food compared to forest sites (at least, at certain times of the year), or (4) the tendency for smaller chimpanzee communities to be more cohesive in general. West African chimpanzees have been described by Christophe Boesch and colleagues, who study chimps at Taï National Park in Ivory Coast, as "bisexually bonded," where adult males and females are found together more frequently than in the well-studied East African chimpanzee subspecies in particular. So, some of the cohesiveness we see at Fongoli may be due to this subspecies tendency. However, recent work by Camille Giuliano and colleagues at the Issa, Tanzania, savanna chimpanzee site also reveals that these apes range in relatively large parties, similar to West African chimps.

While the cohesive nature of the Fongoli community is not atypical of the West African chimpanzee subspecies, in general, their more cohesive nature could also be a result of efforts to maintain social ties in a group that ranges over an extremely large area. The home range at Fongoli is more than ten times as large as that of the Gombe chimpanzees in Tanzania, where Jane Goodall began study in the early 1960s. At Gombe, the chimpanzees' small, 8-kilometer home range means they can hear individuals advertise themselves via long calls or pant-hoots from almost any part of their home range, as these vocalizations can be heard from a distance of more than 2 kilometers away. At Fongoli, this is definitely not the case, as the home range edges can be up to 10 kilometers apart. Even the core home range, which chimpanzees use more intensively during the short rainy season, sits more than 7 kilometers from an important secondary ranging area. Individuals may range together more often to keep track of others in their community within such a large space.

The skewed male–female sex ratio at Fongoli, where adult males outnumber females (over average, there were eleven adult males and seven adult females in the group during the sixteen years between 2005 and 2020) could entice adult males to track adult females more closely by associating with them more frequently. Studies of other wild chimpanzee sites suggest that adult female chimpanzees outnumber the adult males in their community. The fact that the opposite is true at Fongoli may influence a number of behaviors here. Certain behaviors may tie in to what appears to be male tolerance as well as female assertiveness at Fongoli, from the frequency of tool-assisted hunting by females to the dominance by adult females in some contexts—like taking tools, food, and termite-fishing spots from adult males. Regardless of the underlying reason, I think the social cohesiveness at Fongoli influences many aspects of their behavior and ecology at this site. But, while the chimpanzees' behavior must be considered within the social context, the environmental context at Fongoli also looms large.

One thing we know for certain: after decades of studying chimpanzees, we see quite a bit of variation in the behavior of this species. Some of this variation differs according to subspecies. Some variation is due to the environment. In the past, primatologists have tended to generalize chimpanzee behavior, largely basing it on the first studies done on chimps, by Jane Goodall in Tanzania, a site where the East African chimpanzee lives. Indeed, most studies of chimpanzees have been conducted on the East African chimpanzee species (*Pan troglodytes schweinfurthii*), with only a handful of long-term studies conducted on the West African chimpanzee (including Fongoli). There has also been a tendency for primatologists to compare and contrast chimpanzees and their sister species, bonobos (*Pan paniscus*). In fact, like much of behavioral ecology, the behaviors we have tended to dichotomize lie along a continuum. Perhaps the best known is that of aggression versus affiliation. Chimpanzees have been popularly described as "fighters," while their close relative the bonobos have been described "lovers" or as "the hippie ape." In tense situations, bonobos often use sex rather than aggression to defuse tension, while chimpanzees are one of the few mammals who engage in relatively frequent lethal coalitionary aggression, where adult males in particular from one community join together and attack lone individuals from other communities, sometimes killing them. (Humans, of course, are another mammal that does engages in this kind of aggression.) Although little information is available on the subspecies of wild central African (*P.t. troglodytes*) or Nigerian (*P.t. vellerosus*) chimpanzees, the data we do have on East African chimps, West African chimps, and bonobos indicates that some of these behaviors are less distinct among chimp and bonobo species than was once thought, but are still emphasized in popular science in particular. Chimpanzees in West Africa engage in significantly less lethal aggression than those in East Africa, resembling bonobos in this aspect. The West African chimps' tendency to associate together more cohesively within a community, rather than split into solely male or female parties, is also more similar to the bonobo than the chimps in East Africa.

SHARING IS CARING?

In 2007, Stacy Lindshield (now professor at Purdue University) and I published a short paper on the plant-food sharing we observed between unrelated Fongoli chimpanzees in particular. Humans are known for our extensive sharing behavior—think about all the meals you've shared with friends and family. And most primates exhibit extensive sharing from mother to offspring. But fewer species show sharing between fathers and offspring (with some key exceptions, like tamarins and marmosets), and exchanging food with nonrelatives is even rarer. It's seen more in bonobos. In chimpanzees, sharing outside of mother–offspring relationship is seen most often when someone has captured a monkey or other vertebrate prey. Even then, active sharing, where individuals actually *hand* another a piece of food is not very common. What some scientists call "tolerated theft" or scrounging is more common among chimpanzees. I think it would be better termed "tolerated taking" or "passive sharing."

In our initial paper on plant-food sharing, we reported a relatively small sample size, but the behavior was striking for all the chimpanzee rules it was breaking. Specifically, we recorded adult females taking plant foods from adult males and also taking their termite-fishing or galago-hunting tools from them, then taking over their fishing or hunting sites. Oddly enough, the foods that females took from males were not especially rare or difficult to get, as has been seen in some chimpanzee populations that share very large foods. There are a number of explanations that could account for the patterns we see at Fongoli, and Dr. Angie Achorn tested several of these hypotheses for her dissertation work at Texas A&M University.

At some chimpanzee study sites, primatologists have found support for reciprocity in the form of exchanging meat for alliances, which can be seen via grooming, for example. In a few cases, there has been support for what has been called the "meat for sex" hypothesis, in that adult males are more likely to share meat with females they then mate with. There is also support for begging effort as well, where some chimps are apparently exceptional in harassing the meat owner so much that the owner just gives over a piece of the spoils to have some peace and quiet already. At Fongoli, older adult female Farafa would definitely be described in the latter category. I've seen adult males with meat get up and flee to avoid Farafa's approach instead of just refusing to share with her. Another persistent beggar in the Fongoli group is adult male K.L.—he is less likely to cause individuals to flee at his approach, but he is dogged in his patience in waiting for a piece of meat.

Angie found support for some of these hypotheses in various forms. She found that males tended to share with adult females they mated with over the short term, but these relationships waned over time. The best explanation for the meat sharing we recorded at Fongoli over the course of a decade was reciprocity: individuals exchanging meat in what looks like a direct give-and-take of a highly prized resource. Since plant-food sharing is not as common as meat sharing, we have been steadily building up our samples of plant-food and tool sharing/exchange/theft. We plan to test these hypotheses for resources other than meat. I think that long-term relationships are key to the patterns we attempt to understand

Figure 2.1. Alpha adult female Tumbo shares monkey meat with son Cy. Photo courtesy of the Fongoli Savanna Chimpanzee Project.

at Fongoli and elsewhere, and females perhaps play a more prominent role in Fongoli sociality than is the case for chimpanzees in East Africa, for example.

MAMA'S BOYS

Another behavior we see at Fongoli is the tendency for females to support their sons in their quests for dominant status. Given the relatively small community size at Fongoli, as well as the relatively short period of time we've been studying them (at least, compared to studies of sixty-plus years at Gombe and Mahale in Tanzania, East Africa), we have only a handful of examples of this behavior thus far. However, at least in some cases, this behavior appears to be especially significant.

One of the best examples of a mother's influence on her son's dominance status at Fongoli is Farafa. Since we follow adult males as focal subjects, we get only glimpses of female behavior; still we see quite a bit, given that the group is so cohesive. Farafa's matriline is one of about ten we've known over the years at Fongoli, and she is one of the most prolific mothers, based on the six offspring we know she has produced. (This is a conservative estimate—she probably has had at least seven babies.)

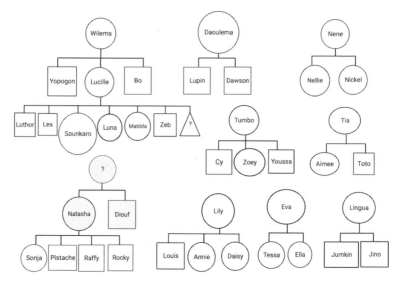

Figure 2.2. Diagram of matrilines at Fongoli from 2001 to 2023, excluding Farafa's (see fig. 2.3) and Nickel's (see fig. 2.4). Circles represent females and squares represent males. By column: **Wilema** (**Lucille** [Luthor, Lex, Sounkaro, Luna, Matilda, Zeb, Kalifou (male)], Yopogon, Bo); unknown mother (**Natasha** [Sonja, Pistache, Raffy, Rocky], Diouf); **Daoulema** (Lupin, Dawson); **Tumbo** (Cy, Zoey, Youssa); **Lily** (Louis, Annie, Daisy); **Eva** (Tessa, Ella); **Nene** (**Nickel** [Teva, Vincent, A.J., Aviv, Kandia], Nellie); **Tia** (Aimee, Toto); **Lingua** (Jumkin, Jino).

When the project was in its early years, there was an adolescent female we could not positively link to either Farafa or another female, but she associated with Farafa. She appeared to transfer out of the community before we could get a fecal sample for DNA analysis of relatedness, but it's likely she was Farafa's daughter, born after Mamadou and before David.

In the case of Farafa, she has always been a well-habituated female, but her involvement in male-dominance struggles was also common, especially when her sons were involved. She came to her older son Mamadou's aid more than once, fighting intensively with males who were attacking him. While females can be injured in fights, they usually receive fewer wounds and less-severe wounding than males do. This doesn't mean that there is no cost to taking up for your son in a dominance battle. I've seen alliances shift in the middle of a fight, and Farafa has always garnered the support of at least a couple of males once she has inserted herself into a dominance contest.

One of these incidents really sticks in my mind, even a decade later. The group was at what we call Petit Oubadji ravine, and adult female Lucille was in estrus. There were usually several males in her vicinity, making sure no one mated with Lucille (and trying to mate with her themselves). In this case, it was middle-ranking adult males Diouf and Bilbo. Audaciously, young adult male David displayed up to the tree they were all in, shaking branches as he climbed and swaying with hair on end—he climbed up, and copulated with Lucille! Bilbo and Diouf immediately began warning calling at David and chased after him. A number of other males also gave chase.

The Petit Oubadji ravine here is narrow, with big boulders along the bottom, so it was impossible for me to keep up with the chimps as they ran back and forth up and down the ravine. Farafa had jumped into the fracas as soon as she heard David and, at one point, she chased after Bilbo, with most of the adult males running after her, some of them in support and others perhaps against "Team David." They all ran out of sight. I stood there wondering whether I should climb up out of the ravine and skirt around to try and catch up with them or just stand there in hopes that they would all run back my way yet again, when I heard a chimp in the bushes. I turned around and David stuck his head out. I became kind of annoyed, in fact. Here was his mother, an elderly female chimp, chasing after adult males in his defense, and he was *hiding*! This is only one example of Farafa's aid in helping David ascend the dominance hierarchy and maintain his alpha status. If she was around and David was in a scuffle, she was there and had his back.

The current alpha female, Tumbo, is the mother of young adult male Cy, who has ascended the male-dominance hierarchy at an astounding pace. Cy became alpha male at the age of fourteen, which is practically unheard of for chimpanzees, since males are usually considered to be an adult socially at around age fifteen (even though they are able to reproduce at a much earlier age). Even becoming an alpha as an older teenager, like David did (at around the ages of seventeen to nineteen), is on the young side as far as the average age of alpha male chimpanzees go, which is in the midtwenties at some study sites. As of 2023, Cy has a strong ally in fourth-ranked adult male Lex, but mother Tumbo is arguably as important an ally in this young alpha's tenure as the highest-ranking male among ten other adult males at Fongoli.

Being alpha male can bring great gains to an individual, like mating

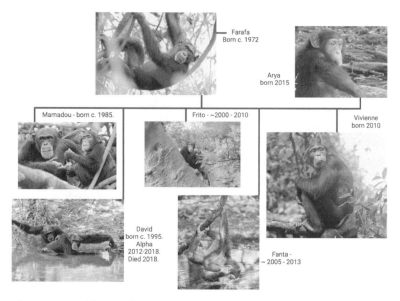

Figure 2.3. Farafa's matriline. Farafa is estimated to be the oldest female in the Fongoli community (born around 1972), and she has had at least six offspring: three sons (Mamadou, David, Frito) and three daughters (Fanta, Vivienne, Arya). Arya appears to be her last birth. Farafa was alpha female during the initial years of the Fongoli Savanna Chimpanzee Project, until Tumbo took over that position. Photos of Farafa and Vivienne by McKensey Miller. Photo of Fanta © Frans Lanting | Lanting.com. Photos of Mamadou, David, Frito, and Arya by the author.

opportunities and priority of access to foods, but it can also come with a lot of costs.

THE COSTS OF RISING IN THE RANKS

Chimpanzees have been described as xenophobic, meaning "fearful of strangers," and I believe this is an accurate description on some levels. Such xenophobia is expressed in the form of aggression, sometimes lethal, directed toward members of other chimpanzee communities. At almost every site where chimpanzees have been studied over the long term, adult males engage in a behavior called *boundary patrolling*. This type of reconnaissance is done at the edges of their home range or territory, and males exhibit distinct behaviors during boundary patrolling: maintaining silence (not vocalizing, and also moving very quietly), listening for chimpanzees from neighboring groups, and "sneaking up" on individuals that they find alone or in small parties to attack them. If they do find a "strange" chimp, they may attack him or her, sometimes lethally. When attacks happen, there are almost always many more chimps in the patrol than are found and attacked—males are apparently unwilling to attack other chimps when they do not outnumber them by four or five times the number of individuals in their own subgroup.

However, Fongoli seems to be an exception to this chimpanzee behavior. We have rarely seen anything that resembles boundary patrolling, which isn't too surprising, given the extremely large home range of the Fongoli chimps as well as the low population density of chimps in Senegal savannas. Patrolling the boundary of a hundred-kilometer home range in an area where the population density of chimps is about one per every 5 kilometers takes a lot of energy; it also means a patrolling chimp could rarely be expected to encounter members of other communities.

We have recorded less than ten intercommunity interactions in the twenty years of behavioral observation at Fongoli. These interactions consisted of the Fongoli chimps moving deep into the neighboring "Bantan" chimpanzee community's home range, to about 3 kilometers or so, and having vocal battles with them. Such interactions are skewed in that the Bantan community is not habituated to human observers, so our presence alongside the Fongoli chimpanzees could help them "win" the small number of encounters we have recorded. However, we are usually lagging more than a hundred meters behind the rapidly moving Fongoli chimps in these cases. In the most recent inter community interaction, head researcher Michel Sadiakho reported that the interaction between the Bantan and Fongoli chimps continued after dark. And, while he caught a glimpse of the Bantan chimps during the encounter, they quickly fled. This could have been because of the Fongoli chimps, of course, but the presence of humans must factor into the Bantan chimpanzees' behavior as well during these encounters.

Based on nest surveys and some encounters with the Bantan community, we believe it's smaller than the Fongoli community, and that the Bantan community withdrew in the face of greater numbers of Fongoli chimps. In 2017, this seemed to have been the case, when head researcher Michel Sadiakho and I independently followed some of the Fongoli chimps about 3 kilometers from what we have defined as the edge of their home range into the depths of the Bantan range. The Bantan chimps appeared to keep moving ahead of our group for much of the day. I was eager to see what would eventually happen, but the Fongoli chimps caught a couple of green monkeys and abandoned the intercommunity engagement to have a monkey feast. They did spend the night there, however, and headed back to Fongoli the next day. One of the most interesting observations I made was that adult female Lily, who had emigrated into our group some years before, led the chimps directly to a feeding tree. Her familiarity with the landscape supports the hypothesis that she and likely other emigrating female chimps originally came from the Bantan community.

Given that boundary patrolling is the norm at other chimpanzee study sites, I have always kept an eye out for similar behaviors. Early in the study, there were times when I wondered if the chimps were about to go on a boundary patrol—I remember being with several adult males as they traveled silently, stopping and listening, and proceeding this way for some time. But this was near the core part of their home range, further from the boundary, so it didn't really make sense to me that they would exhibit such behavior here, unless perhaps a neighboring chimp community party was moving within their home range. This has been the case at other study sites, but we have never found evidence for it at Fongoli. Ultimately, such behavior was associated with the detection of a former alpha male Fongoli chimp, Foudouko, who had been ousted and effectively ostracized by the group. Things didn't turn out well for Foudouko, but the within-group aggression characterizing his case may in fact be the norm for the Fongoli community.

FOUDOUKO'S STORY

Originally, I was really excited to hear that Foudouko was sighted about once a year for several years after being ostracized by the adult males in the group. He was usually in the presence of adult male Mamadou or adult female Nickel and her offspring. We had literally written Foudouko off after we hadn't seen him in three months, based on methods used at other sites of recording disappearances as deaths. Chimps who had not been observed for this extent of time were assumed to have died. This is what I thought might be the case after Foudouko was ousted from the group in late 2007, following a huge fight among adult males that also left his close coalitionary partner, Mamadou, injured to the point that we

Figure 2.4. Nickel's matriline. Nickel was first identified in 2002 as a juvenile who traveled with older adult female Nene and Nene's infant daughter Nellie (who transferred out of the Fongoli community at adolescence). Nickel remained in her natal group and has had five infants since her first in 2007. She has the shortest interbirth intervals of any Fongoli mother chimpanzee, and she was long the lowest-ranking adult female (following the disappearance of her mother, Nene). Nickel has given birth to three sons (Vincent, A.J., and Kandia) and two daughters (Teva and Aviv). Photos of Nickel, Vincent, and Aviv by McKensey Miller. Photos of Kandia and A.J. by Carson Black. Photo of Teva by Nicole Wackerly.

thought he would die. Mamadou eventually healed and moved back up in the dominance ranks to his beta position (number two in rank behind the alpha male, where we had recorded him since 2005) within a few years, and we began seeing Foudouko again.

After several years, Foudouko appeared to be trying to insert himself back into the group and the male-dominance hierarchy. We began seeing him every few months, usually in the company of Mamadou and younger brother David, who had become alpha male in early 2012. Mamadou and David were Foudouko's allies, which appeared to aid his efforts to reintegrate himself and to try and dominate many of the adult males in the Fongoli group. In cases where former alpha males are toppled from the hierarchy, they usually remain in the group to reemerge in the lower

ranks. That was exactly the pattern we saw with adult male Lupin, after David ousted him from the alpha position at Fongoli. However,
Foudouko was apparently more ambitious. In 2013, however, his efforts
came to an end, with his untimely death at the hands, feet, and teeth of
his former group and, most likely, some of his own relatives, given that
male chimpanzees remain in the group into which they are born and, at
some sites, have been found to be related to the degree of half brothers.

Often we hear the chimpanzees at night, especially when they nest
at Sakoto ravine, which is only a kilometer or so from our camp at Fangoly village. We can often even recognize the voices of individual chimps
when they vocalize from Sakoto. On this particular night, we heard many
more vocalizations than we normally hear from chimps during the night,
including what sounded like screams and warning barks, and their cries
also indicated that the chimps had left their night nests and were moving
south some distance. The next day, I stayed in camp, as I had come down
with a bout of malaria. Michel came back from following the chimps
early in the morning, before 8 a.m., looking very disturbed. He told me
there was a dead chimpanzee with the group—he thought it might be
Foudouko. (Michel started working for the Fongoli Savanna Chimpanzee Project in 2008, right after Foudouko had been ousted, so he wasn't
that familiar with what the former alpha looked like.) I left camp then
and found the chimps resting and termite fishing in an area that wasn't
far from Sakoto ravine. Foudouko lay on the ground, dead.

When we first walked up on the chimps, no one was really around
Foudouko's body. However, over the course of the next two hours, most
of the Fongoli chimps, especially the males, proceeded to attack the former alpha's lifeless body. Some of the females even tore off pieces of his
flesh and ate them, including Farafa, mother of Foudouko's allies, and
Tumbo. All the males except alpha male David were seen to attack or
display with the body of Foudouko. But adult male Mamadou, in his display, appeared to be trying to rouse his former ally: he dragged the body
by one limb and pant-hooted in Foudouko's face. The other adult males,
along with some of the females and immature individuals, bit and beat
the body, using sticks in some cases. After adolescent male Luthor warning barked at a wound on Foudouko's foot—possibly because he saw ten-
dons moving as young adult male Mike pulled at the foot—most of the
males returned and increased their attack on the body. Young adult male

Bo even bent one of Foudouko's knees backward, breaking the joint audibly. The treatment of the former alpha male by his group members was hard to watch but not surprising. Foudouko, like later alpha male and coalitionary partner David, was an aggressive leader and evoked a lot of fear from younger males especially. Although we were surprised about the demise of Foudouko's body at the hands of his group, we were less surprised when alpha male David met a similar fate five years later.

In the months following Foudouko's death, Mamadou often left the social group and apparently ranged on his own for several days at a time. Qualitatively, his behavior appeared subdued in these early weeks. If he were a human, I would say he was depressed. It is possible that Mamadou was eventually expelled from the greater Fongoli social community similar to what was initially observed with Foudouko. In February 2014, Mamadou disappeared from the social group and has not been seen since. He was in good health at the time of his disappearance. David, as well as some of the other males, behaved as if Mamadou were in the vicinity for several months. For instance, David moved to the periphery of the area occupied by his party to extensively pant-hoot and buttress drum. This happened more often near the end of the day, and, on at least one occasion, David appeared to leave the nesting party, perhaps to meet up with Mamadou, as I had once seen him do with Foudouko. I discovered a fresh nest in June 2014 several hundred meters from the main nesting party, which consisted of all other Fongoli chimpanzee community members at Sakoto ravine. This was surely Mamadou's nest.

I did not record other possible signs of Mamadou, however, until July 2015, when the group again behaved as if they heard and subsequently chased another chimpanzee for more than a kilometer. A party of young adult males (Bo, Jumkin, Luthor, Mike) raced toward the south of the home range and eventually encountered the oldest and lowest-ranking adult males (Bandit and Siberut) at the Fongoli streambed. The younger males excitedly alarm called and buttress drummed for several minutes, looking south, while the older males calmly termite-fished and ignored them. It is possible that the older males had been in contact with Mamadou, who then fled the arrival of the younger males. The area was well over a kilometer from the Fongoli chimps' territorial boundary with the Bantan chimps, and if there had been a "strange" chimp in the area, I would not have expected Bandit and Siberut to continue termite fishing calmly, not joining in the outcry of the younger adult males. This edge of the home range was also where we encountered Foudouko when he was ranging on his own.

I would not be surprised if Mamadou, who was one of Foudouko's strongest coalitionary partners, had indeed been ousted from the Fongoli group. We have yet to confirm Mamadou's presence at the periphery of the community; but if this is the case, it appears that the same cohort of young adult males (Bo, Jumkin, Luthor, Mike) that we suspect were involved in Foudouko's demise successfully ostracized Mamadou. We hope that future genetic analyses of fecal samples collected over the years in and around Fongoli might shed light on this idea. Even if Mamadou met a different fate, however, his younger brother David lost his alpha position, and ultimately his life, to these same younger adult males.

DAVID'S STORY

David was one of the most charismatic chimps I have ever known, and he was one of my favorites. When our study began in 2001, David was an older juvenile. His mother, Farafa, was the alpha female of the group and was carrying around then-infant Frito. David became extremely well habituated to us, and I remember one instance in which he seemed almost offended by my clumsy approach to within a meter or two of where he sat in the tall grass. I was taking a returning graduate student around, reminding her of the different individuals in the group, and I didn't see David sitting there. He let me approach to within a meter before fleeing and crying as if in tantrum, like young chimps do when they are angry. It seemed as if he thought I was approaching him so closely on purpose, to scare him or supplant him from the spot in which he sat. Since I had never done anything of the sort before, I can imagine my behavior being interpreted as startling and socially unacceptable, rather than David actually being fearful.

David got over my faux pas very quickly, however, and he remained one of the best-habituated chimps in the community. He seemed fairly laid back to me, until I saw his rise up the dominance ranks as he got older. His elaborate displays and unrelenting punishment caused fear and excitement within the community, but David also provided support to allies and stepped in when the situation needed leadership. Head researcher, Michel Sadiakho said it best in a note he wrote in his data book in 2020. Michel compared David to alpha male Jumkin (who was beta male and inherited the alpha position after David's death in 2018) and wrote that Jumkin did not support other chimps as David did in conflicts. Jumkin often avoided conflicts, in fact. While he was not as aggressive as David was toward some juveniles, for example, Jumkin did not seem to express the same leadership that solidified the group as David did. Michel notes that, for this reason, it is hard to forget David.

Indeed, when I finally got the chance to visit the Fongoli chimpanzees in 2021, after an eighteen-month absence due to the COVID-19 pandemic, I remember leaving the entire community of thirty individuals one evening. Leaving the chimps in their nests, settled down peacefully, at the end of a long and tiring day is a very satisfying feeling. It was in this case as well, but I had the distinct feeling that the group was so much smaller, even though it wasn't. It seemed as if the ghosts of David, as well as some of the other males that had disappeared since his demise, like Bilbo, Bo and Luthor, contributed to an absence of some sort. Maybe Michel and I had just grown used to the dramatic scenes associated with David's reign as alpha for six years.

While six years may not seem an extensive length of time for a male to hold alpha position, this period saw David through from around 18 to 24 years old, early adulthood up to almost beyond his prime as an adult male. After Mamadou disappeared in 2014, the young adult males (Bo, Jumkin, Luthor, Mike) attacked David in concert more than once. Higher frequencies of wounding characterized the community for several months, but the social hierarchy stabilized in late 2014. Despite this stability, in January 2015 and again in early June, David sustained extensive wounds from attacks by multiple adult males. Other males besides David also received many wounds in various instances, and these cases almost always coincided with the presence of an estrous female.

David modified his leadership to include more allies in his social circle after Mamadou disappeared. He began cultivating close relationships with the older adult males. Bilbo and K.L. were strong allies, but they were old enough that they didn't really pose a challenge to David's position as alpha, like several of the younger males could have and did do. If one observed closely, you could see that David had lost a step or two over the years. He did not let this diminish his aggressive alpha-male style; however, what this did mean was that eventually David was unable to escape younger males he had put in their place in previous years. In fact, he had never been as fast as younger adult male Bo, but Bo had always previously deferred or at least not caught up with David purposefully in my observations. Given the large number of adult males in the Fongoli group, being an aggressive alpha is a dangerous position. Even if David's closest allies refrained from a concerted attack, or even tried to aid him, there were up to seven other males in the wings that had probably borne the brunt of his wrath at one time or another. This is what we think happened in early 2018, during the dry season, when Michel noticed that all the adult males in the Fongoli community except David were together with several of the Fongoli females, and many of the males bore wounds. Michel feared that David had been attacked. David had recuperated from what was likely an attack by several of the younger adult males, judging from the nature of his wounds, which must have been made by sharp canines, rather than the blunted ones of older males like Bilbo, Bandit or Siberut. The rest of the males were successful in keeping David from drinking the next day, which can have severe consequences during the dry season. But David bounced back—this time. The following year, when a similar situation seemed to take place, David did not recover.

It was February 2018, and David had been missing for several days. Michel eventually found him with his mother, Farafa, and his younger sisters, Vivienne and Arya. David exhibited so many wounds that Michel said it was difficult to look at him. Most of his digits were missing, and he had a hard time climbing—he fearfully climbed up a tree as several males approached him that had been with the subgroup Michel was following. Farafa—David's mother, now elderly, and who had always been his staunchest defender—launched herself at these younger adult males in defense of her son. They did indeed stop, leaving him to rest. Around ten days after we think David was attacked, however, Michel found him dead in this same place, at the southern edge of the Fongoli range, close to a water source.

Jumkin became alpha male after this, until Cy took over the position in 2022. We had predicted that Luthor might overthrow Jumkin, since Luthor had been one of David's strongest and most persistent challengers during the last years of his life. However, Luthor, along with his slightly older uncle, Bo, disappeared in 2020. Neither appeared ill, but Michel noted that Bo began distancing himself from the group after soliciting support in various agonistic conflicts but not receiving any. This seems to have been the same scenario that characterized adult male

Bilbo's social standing before his disappearance in 2019. While these three adult males could have succumbed to some accident or disease, the case of Foudouko makes us wonder whether these adult males and others, like Mamadou, could have or might still be living at the edges of the Fongoli range.

In the late 1960s, Japanese primatologists who began work in the savanna woodlands of Tanzania hypothesized that chimpanzees in such environments might live in more extended social structures. While I think that chimpanzees in savanna environments do live in closed communities very similar to those of chimps living in more forested habitats, the nature of savannas and the extremely large home ranges and low population densities of chimps there mean it would be easier for lone males or small groups of males to live while avoiding or going undetected by their former groups. As additional communities of chimpanzees living in savannas are habituated to observer presence, we will be able to learn whether Fongoli is typical of these kinds of scenarios or an exception because of the skewed nature of their sex ratio—having so many more males than females, and therefore more competition between males compared to other sites.

One reason it's difficult to come to firm conclusions is because chimps have such long lifespans, not unlike humans. We really need to study a group for decades to even begin to understand them in a larger social and environmental context. But our anecdotes—or case studies—can still give us insight, or at least allow us to hypothesize about the larger patterns of behavior. And while of course we focus on the chimpanzees' lives, the small cases of deaths we have seen or have evidence of are also important in understanding our closest living relatives.

FRITO'S DEATH

Water is a significantly limiting resource at Fongoli, unlike most sites where chimpanzees are studied, even in other savanna sites—at Issa in Tanzania, for example. The case of adolescent male Frito provides some insight into how the semiarid savanna biome can be deadly to chimpanzees, whereas similar circumstances for chimps living in forested environments or even other types of savanna biomes would not lead to such a result.

In contrast to Foudouko's apparent immediate death following an attack by males in his group, we think Frito died from his relatively minor wound days later. When we found Frito's body, during the late dry season in 2010 after he had been missing for ten days, it was partly mummified. Skeletal analyses showed a tooth mark in bone near his elbow, typical of a canine wound—but it was a wound that would not usually be life threatening, according to Dr. Adrienne Zihlman, who has extensive experience studying primate skeletal material. However, given the significant stress associated with the late dry season, it is likely that Frito was unable to travel the 3 kilometers between water sources and succumbed to dehydration exacerbated by blood loss. Supporting this interpretation is our observation that David remained in the vicinity of a water source following the 2016 and 2018 attacks on him. Additionally, a badly decomposed older adolescent male was found near a water source in the late dry season in 2002, before we had identified all individuals.

Researchers discovered Frito while traveling with a party that included his maternal brother David. An unidentified male displayed and dragged the body, which the researcher, Waly Camara, interpreted as an attempt to rouse Frito, not as aggression directed toward an individual. Despite the heat during the late dry season, the party remained in this open woodland habitat, sitting near Frito's body for several hours around midday before continuing another kilometer and a half or so to a water source. No additional wounds were seen on Frito's body or bones besides the wound at his elbow. No individuals other than the one male that dragged Frito's body were seen to disturb his body.

The case of dead adolescent male chimpanzee Frito can be considered an example of an incidental by-product of wounding rather than a coalitionary attack. The apparently concerted attack on alpha male David resulted in his death about ten days later, when he did not recover from his extensive wounds. And Foudouko's death clearly resulted from being attacked. These three examples of deaths are interpreted very differently in scientific circles regarding the discussion of coalitionary lethal aggression. However, they were almost certainly related to the within-group dominance behavior of male chimpanzees at Fongoli, while at least two (David and Frito) must also be viewed in context of the stressful savanna biome in which these chimps live.

· 3 ·
Coping with Savanna Heat

Water is probably the most limiting resource for humans worldwide. This is definitely the case for people living in rural Senegal, and the same can be said for the chimps of Fongoli. During the long dry season here, high temperatures exacerbate water loss and make Fongoli a dangerous landscape.

One memory that brings this point home is when adult male Fongoli chimpanzee K.L. couldn't get to a water source he needed. A film crew was accompanying me and the chimps that day. My protocol when visitors are out with the chimpanzees entails having them stick very close to me or one of the research team. The chimps accept strangers in their midst if these strangers are associated with one of us humans, but they grow nervous if these strangers wander away from us. On this particular day, the man shadowing me asked if he could go sit down in the shade instead of standing with me in the sun, where I was observing my subject. Heat stroke and heat exhaustion are especially dangerous at Fongoli for people who are not accustomed to the climate, so I relaxed my protocol for his sake, and he went and sat on a fallen log about 10 meters from me, in the shade. I was taking data, so I really only watched from the corner of my eye as he went to sit down, and it wasn't until K.L. arrived that I realized where this man had gone exactly.

K.L. was walking toward a nearby water hole—it was literally a hole in the ground where water drained from a woodland area into a depression, which eventually flowed into a nearby seasonal stream. This was during a time of year when the stream wasn't flowing, but the deep hole still had water in it from the last rain. I was watching K.L. out of the corner of my eye, but then he stopped directly in front of me and pointedly looked in my direction. It took me a second or two to realize why he had stopped his beeline approach to the water hole to turn and look so pointedly at me (he was definitely looking at me—there were no chimps or other people around me). I realized that the unfamiliar man who had been next to me earlier was now sitting too close to the water source for K.L.'s comfort. I turned and waved at the man to move back, which he did, and K.L. immediately continued on to the water hole and drank. I felt really embarrassed that I hadn't noticed that the man sat too close to the water, and I was a little surprised that K.L. so clearly identified me as the one who could remedy the situation for him. I'm not certain that his association of me with that person was due to the affiliation that he interpreted we had or the observation that I sometimes boss film crews around. Probably the latter. In fact, the chimps might think of me as "that old, slow female who can't run from bees but sometimes chastises men with big cameras." K.L. didn't try to communicate with the man—perhaps because he wasn't comfortable with him, or perhaps he can see that almost all the "strange" humans start out with a lack of understanding when it comes to chimp communication. At any rate, I still feel guilty when I think about that look K.L. gave me. But I feel very privileged to be in situations where wild chimpanzees can communicate with me with a glance. (Okay, more like a hard glance. Maybe a stare.)

The fact is, water is very serious business to chimpanzees in Senegal. If the water sources were not available—because of climate change or human encroachment—the chimps wouldn't last long, as they must drink almost every day, especially during the dry season. Even though there may be water sources available during the long dry season, which lasts about eight months, these become smaller as the season goes on. Ultimately, the chimps are left with only a few sources of drinking water at the peak of the dry season, in late March through early May.

I've mentioned before: anyone who has worked at Fongoli can tell you it's a hot place. It might be the humidity for a couple of months, but it's really the heat for many of the dry season months. For much of the year, there is no rain, and temperatures can be extremely hot, exceeding 43.3°C (110°F). Humans are better adapted than other great apes to living with high heat stresses, even excluding our cultural adaptations like air conditioning or fans. Although humans need to drink daily, we sweat more than other apes, which helps keep us cool. Our relatively hairless body helps in that regard too. Even our bipedal posture and locomotion reduces the amount of the sun's rays that hit us full on at a 90-degree angle. Carrying our body above the ground might give us access to cooling breezes as well, which are especially effective on a sweaty, hairless hominin body.

Even though the chimps at Fongoli appear well adjusted to their savanna environment, the water scarcity and high temperatures create selective pressures that shape their behavior for much of the year. Coping with heat stress and lack of water is part of any animal's challenges when regulating their internal body temperature so they can function under various environmental conditions.

In mammals like primates, *homeothermy* refers to this regulation, and thermoregulation refers to temperature regulation in light of such challenges. Thermoregulation has long been heralded as a potentially significant selective pressure for early hominins (the earliest members of the human lineage) and has been used to explain many of the uniquely human traits I just mentioned: hairlessness, sweating, and bipedality. I use the Fongoli chimpanzees to provide some insight into the kinds of stresses early hominins might have faced in a similar environment, since more closely related species are likely to experience and respond to pressures in similar ways than more distantly related ones (like monkeys or other animals).

Chimps at Fongoli provide some of the first observational data on apes in a hot and dry environment like that inhabited by our ancestors and other relatives within our evolutionary family who lived in a savanna woodland millions of years ago. During the dry season when most trees at Fongoli lose their leaves, pressures associated with UV radiation are significantly higher since late-dry-season temperatures routinely exceed 40°C (104°F), which is above the thermal neutral zone for chimpanzees (17–29°C or 62–84°F) and above their mean core body temperature of 37.25°C (99°F). Unlike research on other primates, such as baboons, the endangered or critically endangered status of chimpanzees limits the use of invasive methods, like implanted telemeters, which measure body temperature and other physiological traits and provide direct data on animals' biological responses to stresses like heat. Instead, we use indirect measures, such as heat-sensitive cameras, to collect thermoimaging data that has provided us a noninvasive yet reliable means of assessing stresses associated with the savanna environment.

Using thermoimaging technology, we have tested hypotheses related to the early hominin niche but also have explored how the savanna biome at Fongoli may change with the increasingly warming climate and how it will ultimately affect the chimps. McKensey Miller conducted her master's thesis research at Fongoli while a student at Texas State University, and she focused on better understanding the chimps' use of different habitats and microhabitats by assessing their activity in areas like forests versus grasslands and under shade versus full sun. She used a thermoimaging camera to record the temperatures the chimps experienced under these various conditions. McKensey found that the environmental temperatures that chimpanzees experience, as measured by thermoimaging their dorsal surface, or back, differed according to habitat—not surprising, given the relative coolness of the small patches of shady evergreen forest and even the woodlands compared to the open grasslands. The chimpanzees also regulate and lower their surface temperatures by resting more and traveling less. However, she found significant differences between individual male chimpanzees and the temperatures they were exposed to, with the alpha male Jumkin and the two lowest-ranking and oldest males, Bandit and Siberut, recording the hottest body-surface temperatures. Behaviorally, perhaps the older chimps did not have access to the same select shady spots, which Fongoli chimpanzees compete over. As far as Jumkin goes, however, his higher body temperature may reflect the stress and pressure associated with the alpha position. Stress-hormone studies at other chimpanzee study sites have shown that alpha males have higher cortisol levels than subordinate males, and various studies conducted at Fongoli have shown that alpha males move more than subordinate males, suggesting they spend energy to maintain their dominance status. McKensey used aspects of the Fongoli chimpanzees' behavior and ecology, as well as data on captive chimpanzee biology, to model the predicted niche of the Fongoli chimps as climate change influences this area of Senegal. Based on her estimates—which take into account the heat and water stresses felt by these chimps, as well as chimpanzees' thermal limits and metabolism—the Fongoli community may not be able to survive in this area of Senegal beyond the year 2080.

If the several permanent water sources available to the Fongoli chimpanzees during the peak of the dry season somehow disappear or otherwise lose their potability, the apes here might disappear much more quickly. Drinking is such a key behavior during this time of year that we, as researchers, make a note of days when the chimps we follow do *not* drink. For lactating mother chimps, skipping a day of drinking is even rarer, and the survival of nursing offspring likely depends on their mothers' almost-daily drinking schedule during the late dry season. As I mentioned in chapter 2, the deaths of certain individuals, like adolescent male Frito, even seem linked to dehydration during the late dry season following an otherwise nonlethal wound. For the moment, though, the Fongoli chimps still have access to this all-important resource, and certain aspects of their behavior strongly demonstrate how significant water is to chimps in this savanna. And there are a number of ways they appear to try and minimize heat exposure and the compounding and associated (lack of) water stress: resting extensively during the day; moving around and feeding at night during the late-dry-season months; and resting in small caves, where temperatures are significantly cooler.

Another behavior unique to Fongoli chimps is their dedication to soaking in pools of water at the beginning of the rainy season. They typically do so when water becomes available for drinking with the first rains, but when temperatures have yet to subside and the relative humidity makes the heat index the highest of the annual cycle (often up to 51°C [125°F] or higher). As part of her doctoral research at the Max Planck Institute in Germany, Erin Wessling extracted hormones from urine samples that she collected from the Fongoli chimpanzees, after waiting for them to move away from the spot where they had urinated and then pipetting it from the ground or leaves into vials. At times, Erin had to wait so long for the chimpanzee to move away that the urine had dried up. She was able to collect a good number of samples, however, and she then examined physiological responses to diet and water stresses in her analyses. The hormone levels of Fongoli chimpanzees showed that they remain stressed despite the various behaviors they use to adjust to their environmental conditions. Erin's results confirmed our behavioral observations of chimps at Fongoli and their reactions to heat and associated factors like water availability as a major stress in their lives on the savanna. Thus, a key stressor associated with the semiarid Fongoli environment is dehydration, and this is likely the limiting factor for chimps at the edges of their species' geographical range in northwestern Africa, rather than food availability or other factors.

Figure 3.1. Adult male Bilbo carries ripe baobab fruit during the dry season at Fongoli. Photo courtesy of the Fongoli Savanna Chimpanzee Project.

AQUATIC APES

During the transition from the end of the hot dry season to the early hot rainy season, when the relative humidity is so high, we often find the whole community of Fongoli chimpanzees together near Sakoto ravine—the site of what seems to be their favorite soaking pool and of Sakoto cave, also a cool place to hang out. While the chimpanzees soak in pools formed by water runoff at the edge of gallery forests or in small streams, they avoid soaking in the larger Gambia River (with the exception of Cy, when he was a juvenile), likely because of the threat of crocodiles and hippos there (see chapter 5). We placed camera traps at Sakoto pool to capture images of chimpanzees soaking at night. From these videos, we were able to identify individuals and found that the chimps exited their

Figure 3.2. Adolescent males Cy and Pistache soaking in water during the rainy season in 2019. Photo by McKensey Miller.

night nests and soaked for several minutes multiple times a night during the early rainy season.

Sakoto ravine is the product of eons of water action. Rushing water that streams down a long, open laterite pan during the rainy season created Sakoto ravine and pool, which is dry for almost half of the year but fills quickly from the water runoff with the first rains during the early wet season. The rain allows pools of water like Sakoto to fill, and it also provides scattered drinking pools across the savanna landscape, as the bare rocks on the laterite pans provide plenty of small pools for drinking. The availability of drinking water at this time of year allows the chimps to range around in larger groups again, as they aren't constricted to only a few water sources when finding food within a close radius of these sources. One of their top foods, the *Saba senegalensis* fruit, is also ripening during the start of the rainy season. It's found in large patches throughout their home range, so this is another reason we see most of the chimps together at this time.

Often, there are several hard rains in May, but there may then be a dry spell for almost two weeks, so the chimps stay localized around Sakoto pool, drinking and doing some soaking daily. The water level is not very high in the pool—about knee deep on a person—and the water is pretty dirty at first, as the leaves and other rotting detritus that have gathered there since the pool was full of water some six months previous has not yet been washed out. On average, a soaking bout lasts approximately three minutes before the individual moves off to rest and dry off before returning to the pool. The chimps repeat these dips in the pool dozens of times a day. Still, a few diehard water babies like Bandit, an older adult male, sit in the pool for up to twenty minutes at a time, vacating only when a dominant individual comes to get a drink and cool off. Bandit is low ranking and takes every opportunity to grab a spot in Sakoto pool when the more dominant males are not around. His ally, older male Siberut, almost always takes a spot on the edge of the pool and has to hold onto the rock ledge and lower himself down (he usually gives an "aaaaah" type of groan as he sits down, which I think anyone can relate to that has ever eased themselves into a tub). While some chimps may spend more than twenty minutes soaking at one time, they usually take quick dips over and over throughout the day—and even during the night.

Chimps at Fongoli are unique among wild chimpanzees in their use of water to cool themselves like this. For many years, based on what we knew of Gombe chimps in Tanzania, primatologists characterized the species as water fearing, or hydrophobic. While there have since been reports of chimpanzees in sanctuaries and other captive situations cooling off in water, as well as wild chimps wading through water to get access to something, there are no other reported cases where wild apes soak in water extensively for cooling purposes. Fongoli chimpanzees actually compete over prime soaking pools, like the shady Sakoto pool, with females and immatures relegated to the "kiddie pools," as I call them: shallower pools of water that are out in the sun on the plateau. They exhibit this behavior early in the rainy season, then cease as temperatures cool down a couple of months later. Around the middle and later part of the rainy season, we rarely see this soaking behavior. In fact, chimps go out of their way to leap small streams without getting their feet wet—such a distinctly different way of reacting to water compared to the hotter months of the year. Most of the year, there is no water to avoid in these seasonal streams, but, when the streams dry up completely, the Fongoli chimps can still access water just under the surface by digging wells with their hands.

Well digging has been seen in other animals, including warthogs, which also live at Fongoli. Other chimpanzees living in savanna environments dig wells for water too, like those at Toro Semliki Wildlife Reserve in Uganda, where the behavior was initially described by Kevin Hunt of Indiana University. Fongoli provides the only systematic behavioral observations of this activity so far, though. Well digging occurs in dry streambeds and includes the same digging spots year after year. It is also seen adjacent to stagnant water pools near the end of the rainy season, as if the placement of wells is to filter water in these cases. While it seems a simple task, I have seen older adult female Farafa digging a well while not only her infant Fanta waited to drink from it, but juvenile Frito as well. Although younger chimps also dig wells, the fact that they often use wells dug out by their mothers and other older individuals indicates that there is some degree of learning necessary to dig wells for water, or perhaps they are dependent on their mother to find the right location to dig.

Fongoli chimpanzees, like most chimps, show a hand preference, similar to humans being right or left handed, and this is reflected in their well-digging behavior. There are also different techniques to well digging. The classic method involves scooping sand toward the body using the dominant hand, which crosses over the body in front of the individual and pulls the sand toward the body. This is the assumption other primatologists have used to score handedness in unhabituated chimpanzees who exhibit this behavior, like at Semliki, and where scientists have taken an archaeological approach to interpreting evidence left at these wells. However, at Fongoli, we often see individuals switch hands and scoop away from the body, as well, often during the same bout, which would provide a false signature of handedness.

CAVE-USING *TROGLODYTES*

Digging wells and soaking in water are two ways the Fongoli chimps deal with the stresses of a hot and dry environment, but another method they use can be gleaned from the Latin name for chimpanzees. The scientific name for the species, *troglodytes*, literally translates to "cave dwellers," although when German naturalist Johann Friedrich Blumenbach chose this term, he meant it in more of a derogatory fashion, likening chimps to brutish, mythical cave dwellers. (The genus name for chimpanzees, *Pan*, is after the Greek god for nature and wilderness.) During the first year of my study at Fongoli, in 2001, I found out that Blumenbach was ironically prescient in this regard: Senegalese guide Mbouly Camara told me the chimps could be found in Sakoto cave, which I thought was almost too good to be true, since it is so reminiscent of how early hominins or humans may have used caves as the first dwellings.

Using caves coincides with high daytime temperatures, when caves are cooler and stable in temperature compared to other habitat types at Fongoli, like the open grasslands. Chimps also compete over access to caves as resting spots. Certain individuals in particular—for example, lactating females—may use these caves as home bases during the dry season, similar to what Dr. Kelly Boyer Ontl observed at another Senegal site, Kharakhena. She found that lactating mothers, who are likely under the greatest stress in terms of experiencing heat and related dehydration, used caves more than other age-sex class. This was the case even though chimps sometimes faced hyenas and leopards using the same cave. While we have evidence of leopards and spotted hyenas using the same caves as chimpanzees at Fongoli, Kelly has camera-trap images of leopards, chimps and spotted hyenas using the same large cave at Kharakhena. At Fongoli, when we have to search for chimps during the dry season, we seek out their water sources, potential feeding sites, and also caves. Older females, like Farafa, are more likely to be found at cave sites, like Kerouani, Maragoundi, or Sakoto.

In Senegal, leopards, spotted hyenas, and porcupines regularly use caves, while we also have camera-trap images of honey badgers, green monkeys, genets, and, of course, bats using Sakoto cave, where the chimps rest periodically. While a number of animals throughout the world use caves, nonhuman primates rarely do. There have been reports of baboons sleeping in caves to keep warm during cold South African nights and langurs using caves as sleeping sites in China, among other cases. Until we reported that Fongoli chimpanzees use caves to rest in and cool off, however, there was only anecdotal information on these great apes' use of caves. Jim Moore had seen chimpanzees in Mali, a West African country bordering Senegal to the east, emerge from caves during a survey he conducted there. Before we habituated the Fongoli chimpanzees, we tried to record evidence of their cave use indirectly via camera traps. I first published data on cave use by Fongoli chimps from mainly indirect evidence, like knuckle prints and feeding remains. Now we have data from direct observations of chimps entering and exiting caves, as well as from camera-trap video of behavior in caves.

The first time I used camera traps in Sakoto cave, the chimps were too shy to enter the cave. After they became habituated to our presence, they also lost fear of the camera we placed at the edge of the cave so we could record who used the cave, when, and for how long. We were able to collect film of social interactions within the cave, such as when young adult male Jumkin chased young male Mike from the cave after female Lily, who was in estrus, entered the cave. The video clearly showed an example of male–male competition over an estrus female. But from the point of view of an observer standing outside the cave, we saw only Lily enter the cave, then flee as Jumkin ran in after her and emerge chasing Mike. While our view at the time allowed us to surmise what had happened, what we could see from the film was that younger adolescent male Luthor was also in the cave. Jumkin charged past Luthor to chase Mike, who was an older adolescent and more of a threat to Jumkin's status as well as his attention to Lily. Luthor just sat and watched the whole spectacle. After Lily, Jumkin, and Mike ran from the cave, he stayed behind, sitting in the coolness. Without the camera-trap data, we would not have this additional piece of detailed information, and it was actually pretty comical to see Luthor just lying there and watching the whole spectacle.

Besides helping us learn more about the Fongoli chimpanzees' behavior in the caves, the camera traps also let us see what other animals used the cave, which would have been very difficult to record otherwise. The first camera trap set up in Sakoto cave recorded 517 thirty-second film clips of animals within the cave: honey badgers, mongooses, porcupines, giant-pouched rats, green monkeys, bats, and genets, among others. One of the most interesting video clips showed a green monkey in the cave the same time as a genet, with the monkey threatening the genet, which either did not understand the monkey or chose to ignore it. The most frequent animal captured on camera was the porcupine—not surprising, since a group of porcupines lives in the deepest recesses of the cave. Chimpanzees were the third most frequently recorded animal in Sakoto cave, representing 12 percent of all images taken. I wonder whether the chimpanzee skull bones and teeth we once found in Sakoto cave were the product of a chimpanzee dying there, or were cached there by the leopard who also uses the cave, or were brought there by the porcupines, who gnaw on bones as other rodents do, to wear down their continuously growing incisors as well as to gain some minerals from the bones.

One thing we do not see is the chimps using caves at night. But they do take advantage of cooler nighttime temperatures as another way to deal with the daytime heat stress.

NOCTURNAL ACTIVITY AT FONGOLI

I believe that Fongoli chimpanzees and other chimps in this part of Senegal are at the limits of the chimpanzee temporal niche. In other words, they live in a climate that puts them at their threshold in terms of the amount of time they can afford to move around and feed during the day because of stresses due to heat and associated dehydration.

Supporting this notion is the fact that the Fongoli chimps rest extensively during the day and move and feed at night during the dry season to an extent not seen in other populations. When I first examined the activity budget of the Fongoli chimpanzees, I was surprised to see that they resembled mountain gorillas in terms of how much they spent resting during the day—over 60 percent of their time! Now, gorillas are infamous among apes for the extensive amount of time they spend resting. This makes sense when you think of their diet of leafy vegetation and the fact that they don't have to go far to fill up on these foods. They then spend much of their time resting and digesting their low-quality (for primates) diet. On the other hand, chimpanzees almost always include ripe fruits in their diet if at all possible—and these are relatively high-quality foods (for primates). Fruits provide more energy than leaves, but they also require more time to find in the environment.

Given that the Fongoli chimpanzees were spending only about 30 percent of their time feeding, I was perplexed at how they were getting enough to eat. Spending so much time resting and so little time feeding is the flip side of what you see in the "average" chimpanzee at other sites. A small part of the feeding-versus-resting ratio could be explained by observational error—in other words, since the chimps were not as well habituated in early years of the project as they were after more years of study, we may have missed some observations of them feeding, since this activity is more difficult to follow than resting. But it is also the case that the Fongoli chimpanzees have expanded their activity into the nocturnal niche, which is usually limited to primates such as galagos and other strepsirrhines that have biological adaptations to being active under low-light conditions, like having better night vision than diurnal primates like chimpanzees.

In 2010, I spent most of my time in Senegal, and so I took advantage of the situation to spend the nights out with the chimps more systematically. For years we'd known that they move around at night and seemed to do this more in the dry season and when there was a full moon. I systematically stayed out overnight a set number of days throughout the year instead of just taking advantage of a good opportunity to spend the night out, as in the past. I spent a total of about forty nights out with the chimpanzees. I hypothesized that I would see more movement at night during the dry season, when the chimps would respond to the intense dry-season temperatures. Much as they use caves during this season as a respite from the heat, I predicted they would intensify their nighttime activity, including travel and feeding, during the hottest times of the year. To test my hypotheses, I would spend three days following the chimps during the daytime, then spend the night of the third day out with them as well. Over the course of the year, I was able to compare their behaviors across seasons and moon phases.

The first night I spent out was with adult male David. I had meant to follow Mamadou, but I lost the sound of Mamadou's footsteps and found David instead. The chimps continued their activity after the sun went down. David hadn't seen me around for a while, I guess, and probably assumed that I had gone back at dark to wherever it is we humans go. So when he heard me, he quickly scampered up a tree a few meters, and I felt pretty guilty about startling him. He just as quickly composed himself, coming right down and moving off about 25 meters to make a nest. I had hoped to position myself at night within this same distance of about half a dozen individuals where I could reliably make out what they were doing. But this was fine. Obviously, I wasn't going to be able to tramp around in the dark and actually find other chimpanzees now—especially after we heard pant-hoots at least half a kilometer away, at the Maragoundi ravine. So, I laid out my nifty bamboo pseudo air mattress (which was to survive a whole eight nights out with the chimps before being popped by the stem of a dry shea butter tree leaf) and stayed with David during a relatively calm night.

(As an aside, that bamboo air mattress was so noisy that one night I'm sure that adult male Bandit climbed down from his nest walked over to a nearby tree and gave a pant-hoot and buttress drum because he was annoyed by the sound. He then returned to his nest, and I didn't move a muscle for ages.)

Another time I slept near David at the Fongoli creekbed. Normally when I spend the night out, I leave right away the next morning to meet up with one of our team, who will follow the chimps that day. On this morning, David arose before light and began eating bamboo. It sounded like he was about a yard from my head, but bamboo is a noisy food, and I guess he was about 10 meters away. It was almost as if he were doing this on purpose. There was plenty of bamboo everywhere—why did he have to munch in my ear so early in the morning? He finally moved off once it began to get light, and as I gathered up my things, I heard him pant-hoot and buttress drum on a taba tree a few hundred meters down the creekbed. This was in the direction of Fangoly village anyway, and I wanted to see if he had met up with anyone. Since chimpanzees are a fission-fusion species, they often join up with other individuals through-out the day, or split off into smaller groups. When I arrived at the tree, I saw that David had met up with older adolescent male Jumkin. As I arrived, David moved off as if I were another chimp he had been wait-ing on—which he and adult male K.L. tended to do more than the other males (wait on us humans). I was actually sorry not to be following him, in case he *had* been waiting on me.

Another morning, after I finished my night of observation, I lay back to wait for Michel to arrive and follow the chimps for that day and I actu-ally dozed off. I was rudely awakened by a pant-hoot and buttress drum that sounded like it was literally behind my head. It was adult male K.L. I imagine he was just advertising himself, as most of the adult males do in the morning—but maybe he thought I needed to get moving along with the other chimps.

I could share an interesting tale for almost each of the forty nights I spent out with the chimps: when a warthog wandered too close and Luthor screamed at it, startling me more than the warthog, which then gnashed its tusks at Luthor; when a bat landed on me; when termites covered me and all my stuff; when Bandit kept jerking on the tree limb my "mosquito hammock" was tied to during the beginning of a thunder-storm, as I tried to pack up my stuff and leave; when I ended up spending *two* nights out in a row because the chimps ranged so far that I knew Mi-chel wouldn't be able to find me—I was incredibly hungry the next day, having no extra food with me, and I was eyeing a ripe *Nauclea* fruit low enough for me to pick, but sweet old male chimp Siberut ate it before I could get to it!

The results of my night studies confirmed our hypotheses about the Fongoli chimps being more active at night and during fuller phases of the moon. But there were also social aspects that influenced the chimps' nocturnal activity—namely, the presence of estrus females. When there was a least one adult female in estrus, there was significant activity at night. I distinctly remember one night when adult female Tumbo was in estrus and seemed to be trying to sleep in her nest. At least three to five adult males sat around below her nesting tree for most of the night, grooming one another or sometimes lying on the ground, as if it were broad daylight.

These active nights make sense of especially lazy days with the Fongoli chimps. I believe that nocturnal behavior (at least, for primates considered diurnal) is remarkably understudied. In part, this is because we humans are essentially diurnal primates as well, and it is more difficult to study animals at night—just ask any primatologist who studies nocturnal species. However, until recently, Western primatologists have also tended to assume that diurnal primates sleep through the night, since this is our cultural notion of what we should strive for in our sleep habits.

All the various behaviors that the Fongoli chimpanzees exhibit to minimize the impact that their hot and dry environment has on their ability to survive draw attention to how important the climate is to these apes. Given that they are already at the northernmost edge of where chimpanzees live in Africa, the abilities of the Fongoli chimps to offset stresses associated with increased temperatures and unpredictable weather events may take them only so far.

· 4 ·

Female *Pan* the Hunter

Tool-assisted or spear hunting is perhaps the most distinctive behavior that sets the Fongoli chimpanzees apart from chimps studied anywhere else. Until we reported our observations, anthropologists used tool-assisted hunting to distinguish our own species from other animals. And at Fongoli, it's the female chimpanzees who exhibit the behavior relatively more often than males. This is the opposite of what you see in other types of chimpanzee hunting behavior, no matter where they are studied, where males dominate hunting behavior. We record, on average, thirty-two such hunts per year and estimate that chimps likely engage in around fifty-five such hunts annually, when we account for the number of observation days we spend with them and adjust the daily rate of hunting to account for days we are not with the chimps. In addition, the average spear-hunting bout is only a few minutes long, so it is likely we are missing bouts as well, since most are shorter. Almost 80 percent of the attempted captures are unsuccessful, and a large number of these are due to the hunt being abandoned, not because the chimp was not able to extract the galago somehow, although that does happen as well. Since adolescent and adult chimps are more likely to abandon an attempted galago hunt much more quickly than juveniles and infants, I assume that the latter are just not as good as detecting that there is *not* a galago in the cavity they are targeting—or that they would be unable to get it because of the nature of the cavity. Observing this type of hunting still excites me, no matter how many times I've already seen it.

It was May 2006, when the dry season is giving way to the rainy season, and the Fongoli chimpanzees start their bush-baby hunting "season."

I distinctly remember watching young adult female chimpanzee Tumbo, who we believe must have migrated into the Fongoli community from a neighboring one, as many female chimpanzees do when they reach adolescence. Tumbo has always been assertive, even though she is one of the smaller females in the group. (Indeed, years later, Tumbo rose the ranks and became alpha female.) She is also one of the most prolific hunters at Fongoli. On this particular day, we were moving through the woodland, and I noticed that Tumbo was walking along, holding a modified tree branch. I immediately thought to myself, "What is Tumbo doing with that tool—where is she going with it?!" She then climbed a nearby tree and proceeded to jab the tool into a hollow in the trunk. This kind of stabbing into tree cavities—in the trunk or hollow branches—is typical of spear hunting, and it also makes a very distinct sound, which alerts us to hunting bouts. That day with Tumbo turned out to be my first observation of tool-assisted hunting, although Paco Bertolani, who had been managing the Fongoli Project for me in 2005, had already observed it several times. He had described the behavior to me and we had wondered what the chimps were after, but we didn't really know until Paco observed Tumbo capture a bush baby later that same year.

Before this observation, using tools to hunt is something we had associated only with humans, and the fact that females lead the way in this behavior also stood out. This particular behavior is what made the Fongoli chimps famous. The prey that the chimps hunt with tools is also different from the prey that most chimpanzees focus on at forest sites, which is almost always the red colobus monkey. The bush babies, or galagos, spend their days nesting in hollow branches or trees at Fongoli, and this is where the chimps try to access them, using the stick tools they construct. The Fongoli chimps almost always use a branch they have taken from a tree near the galagos' sleeping site, and they prefer using a wolo (*Terminalia*) branch if one is close. They strip off all the side branches and leaves and often break off the terminal end of the branch, removing the flimsy tip. Some chimps go even further and trim the tip with their teeth, effectively sharpening it. Only a few individuals, like Tumbo and some other females, remove all the bark from the branch tool. They then stab and jab into the tree cavity with the resulting spear, trying to injure or rouse the galago within so they can either grab it when it tries to escape or capture and kill it without being bitten. I thought removing all the bark from the tool seemed like overkill until I tried it that way myself—

removing the bark also removes a lot of friction during stabbing into tree and branch holes that have various bits of bark sticking out.

That year, when I first observed the behavior, I racked up another dozen observations, and we published our initial paper describing the behavior based on twenty-two cases. Since then, we've recorded scores of such hunts each year, and have now recorded almost six hundred such hunting events. In this chapter, I'll talk more about who hunts what, and how often, based on fifteen years' worth of observations. Like most chimpanzee behavior, individual differences exist, but there are major patterns also. I'll describe various aspects of tool-assisted and other types of hunting at Fongoli, such as how tools are made, as well as how the bush babies themselves shape the chimpanzees' hunting behavior.

Although the Fongoli chimpanzees are currently the only ape community observed to regularly use tools to hunt, it is undoubtedly one of numerous undiscovered behaviors which chimpanzees and other animals exhibit that we will never know about because of our destruction of their habitats. Communities of chimpanzees and groups or other individuals that live at the edges of their species' range are more likely to exhibit behaviors or adaptations to challenges associated with these narrower niches, and they are the ones most likely to disappear first.

WHY STUDY HUNTING BY CHIMPANZEES?

Although a young female chimpanzee at Mahale Mountains National Park, Tanzania, had been observed using a stick tool to hunt a squirrel in the 1990s, it wasn't until we reported multiple cases of tool-assisted hunting at Fongoli that anthropologists yet again revised a definition of our own species. Until this time, it was thought that only humans used tools to hunt other mammals. The "man the hunter" paradigm in various forms had characterized anthropology for almost seven decades. And, the fact that our closest living relatives, chimpanzees, hunt and eat meat contributes to the heavy use of these apes as models to provide insight into how hunting may have looked in our own distant relatives or even ancestors. Additionally, as I mentioned, adult and adolescent females are the most prolific spear hunters at Fongoli, which is the opposite of what has been observed at all other study sites where these apes hunt mammals. As archaeologist Travis Pickering points out in his book, *Rough and Tumble*, on aggression and the evolution of hunting in humans, the emotional control that chimpanzees at Fongoli use in tool-assisted hunting should be of great interest to anthropologists. Pickering suggests that a hunter who can remove themselves from direct physical contact during hunting, like using a tool rather than their hands, is ultimately what would characterize our own hunting behavior as humans. According to Pickering, it also suggests that hunting becomes more measured and less emotionally charged than during the typical type of chimpanzee hunting, where prey is caught and killed without the use of tools. To Pickering, this also means that behaviors and physiological responses associated with aggression would be reduced.

It is easy to make analogies between what we see a living primate doing and what we think that our own ancestors may have done. The key is to construct a reasoned hypothesis for such analogies. A simple example would be a zoo visitor watching apes hitting each other with sticks and thinking this must be what our ancestors looked like before they became "civilized." Comparing the behavior of living lemurs, monkeys, and apes to what we think may have happened in our own past is the most simplistic form of a model and also the least insightful, as far as understanding extinct human or hominin behavior.

Chimpanzees and their sister species bonobos have been evolving as long as we have, and some scholars maintain that chimpanzees are in fact more derived, or changed, than our own species. Genetically, this may be the case; but it is easier to argue that our own species is more derived in terms of some aspects of anatomy and morphology, such as bipedalism and brain size, and the behavioral changes that have come about as a result. At any rate, scientists generally acknowledge such changes in evolutionary history but perhaps often do not go far enough in qualifying their interpretations or explaining any variables that would affect them.

Another issue altogether is the fact that reasonable scientific interpretations are dramatized in the media. More than three hundred news sources reported on our first paper on tool-assisted hunting by the chimpanzees at Fongoli. While most articles were reasonable, the headlines of these articles (rarely penned by the author) could be fairly dramatic, as you might expect clickbait to be. You can imagine what someone might have come away from after reading the title "Chimpanzees Snack on Skewered Bush Babies" without actually reading the well-written article that focused on details of the chimpanzees' hunting behavior, never suggesting that galagos were skewered, not to mention our scientific article upon which it was based. Regardless, scientists should be responsible for communicating their work accurately, and part of this responsibility should entail trying to prevent misinterpretations and misdirections. Using living chimpanzees and bonobos as an analogy for human behavior is rife for such misunderstanding.

The most useful type of model in which chimpanzees and bonobos can be used, in my opinion, has been termed a *relational model*. Anthropologist Jim Moore of the University of California San Diego has written extensively on the role of chimpanzees as models for better understanding human evolution, with an emphasis on savanna chimpanzees in this role. Moore's relational form of a model is one I use in my research on the Fongoli chimps. This involves understanding savanna chimpanzee behavior in light of what we know about these apes living in forested environments—which is to say, almost all other chimpanzee studies. On a very simple level, we might say any differences between the Fongoli chimps and forest chimps is probably related to the different environments in which they live. (In later chapters, I will talk more about some of these differences.) However, the more we learn about chimpanzees, the more we realize they are a very diverse species in terms of behavior. Not all these differences are readily linked to ecological differences between the groups; certain scholars consider these differences to be the result of "culture." The roles of geography and genetics are not yet completely clear, however, in terms of what is driving the behavioral diversity seen in chimpanzees living across Africa. Teasing apart the variables is a goal of the relational approach. For example, in addition to taking into account the possible effects that a savanna habitat has on Fongoli chimpanzee behavior, I also have to take into account their geographic and indeed genetic or subspecies status.

In examining the behavioral differences between chimpanzees living in different habitats (and assuming physical differences are not significant among individuals belonging to the same species), we ideally want to tease out those differences that are linked most closely to environment. The most important pressures affecting living great apes in a savanna environment, for example, are likely to have been the same ones affecting our own distant relatives and ancestors in this type of habitat. Their responses are also likely similar to the ones we see living great apes make today. We expect that organisms more closely related to us, such as chimpanzees, are more likely to exhibit similar adaptations to a common problem than, say, a more distantly related organism like a monkey. While to me these arguments justify using chimpanzees as a type of model to hypothesize about our distant ancestors, I am equally intrigued by what it says about variation in chimpanzees living today. Like many of the differences we see between Fongoli chimps and those living in more forested habitats, the environment plays a big part in shaping behavior.

TYPICAL CHIMPANZEE HUNTING

At almost every site across Africa where chimpanzees have been studied for at least a decade, they have been seen hunting and sharing meat, which is thought to be a highly prized resource based on their behavior. Group members crowd around the captor or owner and beg for even a small piece of meat. At a few study sites, females in estrus actually gain a bigger portion of the meat from a captor, who is usually a male, than do other "beggars." A lot of meat transfer is passive, where other chimps are allowed to take meat from the captor as he eats, or they scrounge scraps that are dropped on the ground. At any rate, even though meat usually accounts for less than 10 percent of items in the yearly diet of chimpanzees anywhere, it does seem to be a highly desired food, and recent studies have shown that even scraps of meat scrounged from the ground can provide essential nutrients.

At most sites where chimpanzees have been studied over the long term, monkey species are among their most common prey. In particular, red colobus monkeys seem to be the favorite prey of chimpanzees who live in the same environments as these large monkeys. At some sites, larger male colobus monkeys try to protect their group members from chimpanzees, who often target the smaller females and immature monkeys. Multiple adult male chimpanzees join in what are sometimes fantastic aerial combats high above the ground, but some also wait below to catch monkeys who have fallen or are thrown down. In West Africa, at the Taï National Park site, primatologists describe monkey-hunting behavior by adult male chimpanzees as cooperative, not just coordinated, behavior. The forest habitat and the behavior of the monkeys themselves influences the hunting styles of chimpanzees at various sites. But monkeys are consistently the main target of hunting chimpanzees across Africa.

Another characteristic of chimpanzee hunting is that it is usually the purview of adult males. At some sites, like Mahale in Tanzania, females may account for about a third of all hunting observed, but this value is usually lower. Females get meat from males, and if a female does capture prey, it is highly likely that her prize will be taken from her by a male. The same is often true for subordinate adult males—their capture is often stolen by a more dominant male, and especially by the alpha male of the group.

WHAT IS HUNTING LIKE AT FONGOLI?

There are a number of differences between the hunting behavior of chimpanzees at Fongoli compared to those at other sites. While males at Fongoli also employ the typical "chase and grab" style of capturing monkeys seen elsewhere, the prey species differ. Additionally, both male and female chimps at Fongoli frequently use tools to hunt. Other primates, like monkeys, make up 40 percent of the prey Fongoli chimps capture, while we have seen a handful of cases of the chimps capturing bushbuck fawns, and, in one case, a female captured and ate a banded mongoose. (We've otherwise only ever seen chimps capture and play with young mongooses, which doesn't ever turn out well for the mongoose—but they aren't eaten, as far as we've seen.) Overall, primates account for 98 percent of the vertebrate prey captured by Fongoli chimpanzees. Almost 60 percent of these prey are the galago. While some chimps—especially males like old Bandit—run down galagos after another chimp has flushed them from their sleeping cavity in a hollow tree branch or trunk, most galago hunting is done via tool use.

Bandit was, in fact, the last holdout as far as tool-assisted hunting goes. Years after we had first discovered the chimps hunting with tools, we had recorded every other chimp in the group that was at least of juvenile age to use tools to hunt except for Bandit. Bandit is one of the few chimps who would thrust his hand into a nesting cavity and pull out a squirming bush baby. Finally, in 2018, we saw Bandit make and use a tool to catch one! And although most adult males focus on monkeys as their prey species, Lupin, a former alpha male, also focuses on galagos. Also, while females predominantly hunt galagos, we have observed them capture monkeys, even baboons. Both males and females capture bushbuck fawns in equal numbers.

The fact that the hunting target for Fongoli chimps differs from that of other chimps may largely be due to the relative rarity of monkeys and other prey individuals at Fongoli. The Fongoli chimpanzee meat-diet profile consists of more-typical savanna-dwelling primates—like patas and green monkeys (close relatives of vervets), as well as baboons—in addition to ungulates, like bushbucks. Patas monkeys, for example, live in the same areas as chimpanzees, but these swift primates sometimes fall prey to Fongoli chimps. The key to catching these fast monkeys— which can reach speeds of 15 kilometers per hour when running on the ground—is to stick to the tree crowns. I once saw adult male Bandit, one of the most successful Fongoli hunters, flat out give up a chase of an adolescent patas monkey after it leaped out of a tree and hit the ground running. And, while Fongoli chimps eat baboons and green monkeys, like chimpanzees do elsewhere, the most common monkey prey species, the red colobus, is not on the menu at Fongoli because they are not found in this semiarid environment in Senegal.

Although I may have given you the impression that Fongoli chimps focus on galagos because they are desperate for meat in this environment, we have sometime seen them capture mammals they don't eat. In addition to mongooses, with that one exception I mentioned earlier, the Fongoli chimps capture but don't eat genets, a catlike creature related to the mongoose family. This might be because of their unpleasant scent glands, but we have observed a few of the chimps capture and play with young genets—a scenario that, as with the mongoose, usually did not end well for the genets! One exception was filmed and appeared on an episode of the BBC's *Spy in the Wild*. Adolescent male Lex found a young genet one day and carried it with him for much of the afternoon, retrieving his little "pet" when it tried to crawl away while Lex was termite fishing. Lex patiently brought it back and set it down by his side. When he rested, he lay on his back and sat the baby genet on his chest.

HUNTING OR GATHERING?

Many scientists did not believe that what we initially reported at Fongoli was hunting (maybe they were playing, one critic suggested), and some still argue that this is "gathering" (especially since female chimps are the ones that do it more frequently) or "extractive foraging." I have heard more than once that hunting galagos should be considered gathering because the prey is small, not dangerous and cryptic, or hidden. If adult male chimpanzees were the characteristic galago hunters, I doubt that whether this behavior qualifies as hunting would be an issue. It seems easier to accept female humans as hunters than it does for at least some primatologists to seriously consider female chimpanzees as playing a part of the meat acquisition niche among apes. At any rate, regardless of the size of these prey, chimpanzees are reluctant to be bitten by one. In addition to the lower return on the hunter's efforts and the relatively inexpensive energetic expenditure, the main difference between galago hunting and the typical chase-and-grab monkey hunting has to do with the complexity of making and using tools to acquire concealed prey that is also capable of inflicting injury—regardless of how minor— to the hunter.

In addition to being fixated on the small size of galago prey, critics also compare this type of hunting to termite fishing. To clarify, termite fishing involves using a twig or grass tool to lure soldier termites out of their mound (which bite onto the tool with oversized mandibles), then sticking the tool into your mouth to eat these insects off it. Although galagos are also concealed, they do not bite onto stick tools and allow themselves to be pulled out, like termites do. Bush babies are often aggressive as well as hidden, and may actively avoid the chimpanzee hunter. Plus, a bite from a termite soldier doesn't break the skin—and I've been bitten on the inside of my cheek after trying to termite-fish myself. Thus, termites don't present the same risk for infection that even a bite from a small mammal would.

The case of tool-assisted hunting that I opened this book with illustrates my point. That case involved three different chimpanzees and one very fierce bush baby. I first saw adult female Lucille jabbing a spear into a tree cavity. She was so intent that I was certain there was a bush baby in the cavity. She reluctantly abandoned the attempt after several minutes, and adult male Karamoko tried his hand with a spear next. Even after they had finished their attempts, Lucille and Karamoko stuck around and watched as adolescent male Frito made a spear and gave it a try. A couple of times, Frito actually reached his hand down as if to insert it into the cavity, but he would jerk it back up, and he also eventually abandoned the hunt. I was able to climb up into this tree and look into the cavity—there was indeed a bush baby inside! It had its mouth open in a threat directed toward me, and I could see that it had wounds on its head from the chimpanzees' tools. (For you bush-baby fans, the wounds

I saw looked superficial, so maybe it recovered.) It was not very far down in its cavity, but it appeared to have a sort of a shelf of wood that it could squeeze back under, avoiding the full effort of the chimps' jabbing and stabbing. The size of the opening of the cavity was big enough for Frito to reach in, but he apparently was averse to being bitten, even by such a small animal.

Even though a galago may provide only roughly the meat equivalent of a single taco, the estimated kilograms of meat captured by adult female hunters at Fongoli places two of them (Tumbo and Farafa) in the top ten hunters of the group. Tumbo, who is probably in her late twenties as I write this, is ranked sixth highest in the amount of meat she has been observed to capture during our study, while Farafa is eighth. One former alpha male, David, as well as older chimpanzees Siberut and Bilbo are the top hunters in terms of the amount of meat we've recorded them capturing.

WHY HUNT SUCH SMALL PREY?

Besides hunting, an aspect of meat eating that anthropologists are very much interested in is meat sharing. Primatologists working at Gombe in Tanzania explored what they call the "meat scrap hypothesis," and concluded that even small pieces of meat obtained from a hunter (either given, taken, or scrounged from the ground) can add significantly to the diet of other chimpanzees in the group. A surprising finding from Fongoli is that, although we expected female hunters to share most often with their offspring, we found that they also transferred meat to unrelated group members. It seems that some of the hypotheses used to explain why male chimpanzees hunt and share meat, such as alliance formation, might also apply to female chimps—at least, at Fongoli. This has important implications for how we view the evolutionary history of hunting in our lineage. Maybe the "man the hunter" hypothesis should be retired for good . . .

The case of Farafa's daughter Fanta might give insight into the nutritional importance of even small amounts of meat. She was one of the most persistent beggars from meat owners, especially during the later stages of an undiagnosed illness that ultimately led to her death. One of the secondary conditions discovered during a necropsy of Fanta's body by a veterinarian and scientists from the Institut Pasteur de Dakar was

Figure 4.1. Juvenile female Tessa tries her hand at tool-assisted hunting. Photo by McKensey Miller.

anemia. It may be that Fanta's anemia spurred her to exert so much effort to obtain iron-rich meat by hunting, and we also observed her persistently begging meat from other meat owners, even adult males, whenever there was a hunt. Her aggressive begging from more-dominant individuals is not typical of what we see in young female chimps especially, which also supports the possibility that Fanta was trying to take care of her anemia.

Fanta was also a very prolific hunter, regardless of the fact that we did not record many successes for this young female before she grew into adolescence. One year, she gave me a glimpse of bush-baby hunting my first day back in the field that year. The chimps hadn't gotten very far from where they'd nested overnight at Sakoto ravine, one of their favorite haunts. (And one of our favorites too—it's a mere twenty minutes from Fongoli camp, allowing us to sleep in past 5 a.m. rather than waking up at 3 a.m. to reach the farthest edges of their home range before they wake up!) Fanta fashioned a spear from a live tree branch, trimming off the side branches and leaves and modifying the tip with her teeth. No luck for Fanta that day, even though she was persistent in her hunting and tool-making behaviors, but she accumulated more than twenty-three bouts during her short life. Fanta was observed hunting even more frequently in just a few years' time than adult female Tumbo, who is the most successful galago hunter to date, capturing a bush baby in more than a third of her hunts.

HUNTING WITH TOOLS

We have observed some key steps chimpanzees have commonly used to make hunting tools. Individuals almost always perform the first two steps, which consist of locating a live branch and detaching it from the tree, and then detaching side branches and leaves. Most hunters also break off the more flexible terminal end of the branch, which effectively makes the tip sturdier, and some females in particular may trim the tip of the tool with their teeth. These tips may be trimmed repeatedly during a bout of hunting, as they become blunted during the jabbing process. Certain individuals, like adult female Tumbo, also remove all the bark from the tool. I didn't appreciate the utility of this until a National Geographic film team asked me to make a spear this way and hunt while they filmed it. A tool that has had the bark removed is easier to jab vigorously into a branch or trunk cavity, because it doesn't catch on the various knots and protuberances inside the cavity (and, at least as a human, you don't skin your knuckles as much).

Chimps may construct several hunting tools during a bout at a single galago cavity, but in more than five hundred cases, we have recorded only a few instances where the chimp uses an old tool, found within the cavity itself. In most of these instances, an immature chimpanzee was the hunter. However, reusing a freshly made tool is not unusual, as one individual after another attempts to obtain a galago at a cavity, and there are multiple instances of tool sharing—which is perhaps better termed *tolerated theft*—usually by females from males. I once saw Tumbo, as a subadult female, approach hunting adolescent male Luthor and calmly take the tool he was using out of his grasp and take over the cavity he had been hunting in. This type of tolerated theft is another characteristic of the Fongoli chimpanzee community that is relatively unusual, given that all adult male chimpanzees are dominant to all females, regardless of the female's age and rank, and adolescent males are also dominant to adult females once they reach a certain age. Tumbo and other females, especially Farafa, have also been seen to approach termite-fishing males, take the tools out of their hands, and begin termite fishing in the exact spot they essentially supplant the male from. At the extreme, I saw Farafa use her butt to shove adult male Jumkin out of a spot after taking his tool, apparently because he wasn't vacating the place quickly enough. Male tolerance of female assertiveness is key to females' role in the tool-assisted hunting at Fongoli, in my opinion, and I'll expand on this subject later.

The main prey species of the Fongoli chimpanzees, galagos, are nocturnal. Grace Ellison, a researcher from the UK, found that Fongoli bush babies are almost always strictly cavity nesters, unlike their counterparts where she studies them in Tanzania, who frequently use dense clumps of vegetation in trees. Grace interprets this as a response to predation pressure from the chimpanzees. I have seen only one case where a bush baby was disturbed from a vine nest, which happened during a rainstorm—in fact, the galago may have left a cavity nest during the rain, as they frequently do. I only saw it emerging from a clump of vines. The galago was ultimately captured and eaten by Tumbo, a subadult at the time, after she threatened away at least two adolescent males—Frito and Jumkin—who had begun to pursue it as well. They stopped their chase, and Tumbo was able to grab it after a few moments of chasing it back and forth in the rain. That case sticks in my mind because I had a front-row seat. Tumbo caught the galago and held it by the tail. I can even remember it turning to look up at her, and I just wanted her to get it over with already. Adolescent male Jumkin approached and seemed nervous, giving a little fear grin. Tumbo reassured him by extending her arm to him; then, almost immediately, using the bush baby's tail as a handle, she cracked its head against a rock and started eating it. I immediately regretted my privileged view of this spectacle.

SEX DIFFERENCES IN HUNTING

Besides the fact that Fongoli is the only community of chimpanzees that routinely hunts with tools, the finding that adult and adolescent females here hunt with tools more than expected (given their membership in the group or in subgroups that we record data on) violates the norm of the predominantly adult male chimpanzee hunters at other sites. When I arrived one year in May, head researcher Michel Sadiakho told me that adolescent female Sonja had already tried her hand at galago hunting a couple of times that season before I arrived. (As I mentioned before, May is when the rainy season at Fongoli usually begins, so we consider

it a transitional month). No luck for Sonya yet, but she'd received an un-
happy surprise when she roused a genet during one bout. I expect the
genet gave her a shock because it was larger than a bush baby and would
have had to run out of the cavity past her—unlike a bush baby, which
would have hopped off had it been so lucky to escape. The next day, true
to form for an adolescent female, Sonja decided to hunt again. She spent
quite a lot of time making, using, and discarding four different spears
before she abandoned hunting at the cavity she had been concentrating on. From her behavior, I deduced that she did not detect bush-baby
presence, which is what I believe accounts for most failed hunts. (There
have been times when it appeared clear that a bush baby was present but
could not be captured, as I described in the case of Lucille, Karamoko,
and Frito earlier.)

Different styles of hunting with tools are seen in males and females,
with the latter usually exhibiting more finesse. Some males go the brute-
force route. Although males make slightly longer tools than females, in
general, the differences between the length of tools made by adults and
adolescents of both sexes at Fongoli is not statistically significant, averaging about 70 centimeters in length (a little over 2.5 feet long). This
also is true for the success rate of males and females. While males have a
success rate of about 22 percent, the 20 percent success rate by females is
not statistically different. This might seem like a low rate of success overall, but galago hunting is also a relatively low-cost behavior compared to
other types of mammal hunting. In fact, when you look at the success
rates of various big-cat hunters as well as chimpanzees elsewhere, Fongoli chimp success rate with galago hunting is about on par with individual monkey hunters' success rates among chimpanzees at Taï National
Park, Ivory Coast, and with cheetahs.

As I've explored various aspects of the tool-assisted hunting behavior
by the Fongoli chimpanzees over the years, one thing I have concluded
is that the nature of the social group here as well as the prey type is key
to the relatively large proportion of hunting by these female chimpanzees. I talk more about social behavior and social tolerance at Fongoli in
a later chapter; for now, I will say that it is likely that female chimpanzees
at a number of sites are interested in hunting, but the fact that dominant
individuals—and in chimpanzee society, all adult males are dominant to
all adult females—regularly steal their catch doesn't provide a lot of incentive for females to hunt, at least in the presence of dominant males.

Females at the Mahale study site in Tanzania actually capture a slightly higher proportion of the prey captured by their community than Fongoli females (around 32 percent at Mahale, versus 30 percent of all prey captured at Fongoli). But when it comes to *tool-assisted* hunting, the Fongoli females account for over 45 percent of the bush-baby captures, despite the fact that there are almost twice as many male chimps at Fongoli and that our protocol entails following adult males. Other study sites, like those in Kibale National Park in Uganda and at other Tanzanian chimpanzee sites, report that female chimpanzees account for between about 5 and 20 percent of all prey captured.

It seems that tool-assisted hunting enables female chimpanzees to capture more vertebrate prey than other types of hunting. Is this related to the fact that female chimpanzees seem to be better tool users in some cases? Or is it because the prey item they obtain with tools is not desirable by males? I reject the latter explanation, because males at Fongoli often beg meat from females who have captured a galago. It may also be that females with clinging infants (and almost every Fongoli female has a dependent infant, unless she is so elderly that she no longer cycles reproductively) can engage in tool-assisted hunting with fewer costs and perhaps less competition than in a monkey chase. Male tolerance of carcass possession does support the fact that Fongoli females have incentive to hunt because it is rare for a dominant male chimp at Fongoli to steal a prey item from a lower-ranking male or from a female; theft appears to be the norm at most other chimpanzee study sites.

Theft—where an individual steals captured prey from the captor—is rare at Fongoli. Indeed, we've only seen such theft in about 3 percent of almost five hundred cases of meat ownership that we've recorded. In the handful of cases we've seen where a chimp stole a prey item before any of it could be eaten by the captor, Farafa's son, alpha male David, was the thief three times, and in two cases, another alpha male at the time of the thefts, Lupin, was responsible. David was a very aggressive leader, and it is not surprising that he was able to steal a prey item from a subordinate, given the deference shown him by other chimps in his group. I also witnessed him steal the opportunity of a capture at least twice. This was in the case of monkey hunting, where a faster chimp, young adult male Bo, was on the verge of capturing a monkey when he shrank back from David's approach and David grabbed the prey. In that case, however, it seemed as if David's possession was tenuous. As usual, a number of individuals crowded around, hoping for a handout, begging for a piece of meat. Alpha female Tumbo was at the forefront of these beggars, who otherwise were predominantly males. Tumbo reached over and took hold of one of the monkey's limbs, slowly pulling it toward her, as if to take it from David. He stood up with his hair on end and threatened Tumbo. Rather than back down, Tumbo reacted by screaming angrily at him. Moments later, Tumbo and multiple adult male chimpanzees were chasing him. He ultimately dropped the monkey, and it was retrieved by midranking adult male Diouf. Later, adult male Luthor stole part of the carcass from David's mother, Farafa. She chased him down with infant Arya determinedly hanging onto her mother's side, and Farafa returned with the carcass, demonstrating another example of the outrage associated with theft at Fongoli.

There was another instance that was so distinctive, it makes me think some sort of theft was involved, even though I didn't witness it. I heard someone hunting with tools—the thwack, thwack, thwack sound is the cue that we most commonly use to tune in to spear-hunting bouts. When I arrived, I found that adult female Lucille had captured a bush baby and that her older juvenile daughter, Luna, was throwing a temper tantrum. Chimpanzee tantrums are much as you might expect if you are familiar with them in humans: youngsters cry and scream and literally throw themselves to the ground. As Lucille began eating the galago and sharing with infant Matilda, Luna's younger sister, Luna cried more and more. She continuously appealed to Lucille for some of the galago by holding her hand out to her mother. When Lucille finally offered her a small piece, Luna became enraged, raced out of the tree, and threw herself onto the ground screaming. She then chased any immature chimpanzee in the vicinity—even older chimps she was usually afraid of. She actually looked my way, and for a second I was afraid she was going to take her anger out on me too. Fortunately, she went back to her mother and continued crying. Lucille, by that time, had gone after Luna and met her partway, offering her an entire galago arm. I will never know, of course, but Luna's outraged behavior indicated to me that her mother had either stolen the galago from her or stolen the hunting site from her, and that she was also insulted by the meager portion Lucille first offered her. Erin Wessling, a graduate student who conducted her master's and doctoral research at Fongoli, once observed Farafa take a bush-baby site from her daughter Fanta, but Farafa shared her capture with Fanta. These interactions between hunting females and their offspring are so intriguing to me, and one of the subjects of our ongoing study at Fongoli. Another time, Fanta had captured a galago but didn't seem to know how to kill it. Farafa stepped in and obliged and then shared the meat with Fanta.

Keeping in mind that we're most certainly underestimating female hunting behavior, if we look at the top ten hunters at Fongoli in terms of success, two of them are females: current alpha female Tumbo and former alpha female and now elderly Farafa. Tumbo is the seventh most successful hunter at Fongoli, behind less than half of the adult males we have known in the group since our study began. In fact, 40 percent of successful hunters at Fongoli are females, which reflects their membership in the group more or less (38 percent of the group is adult or adolescent females). At Fongoli, then, it seems that females are not intimidated by males when it comes to hunting. In recent years, the top hunter position shifted, as alpha male David lost the number one spot to old male Siberut. If David had lived longer, he likely would have held on to this number one spot, but he died at a relatively young age (in his twenties). Siberut was an adult male when we began systematic data collection in 2005, and he has been the lowest-ranking male for most of our study, yet he is an excellent hunter and, despite his low status, doesn't lose his catch to more-dominant males. In 2005, David was only about ten years old, a chimp adolescent, so that he was not yet in his prime hunting years. Like other males, David shifted from a focus on tool-assisted hunting when he was an adolescent to targeting monkey prey as he matured. Using tools, therefore, gives young individuals and females access to meat they might not usually be able to access except via sharing or scrounging from others. And both male and female adolescents are rarely the recipient of shared meat from captors, unless it is from their mother. Using rudimentary tools gives even immature chimpanzees the chance to hunt for meat.

THE TYPICAL SPEAR TOOL

The Fongoli chimps are selective about what they use to make the tools they hunt with to some degree, even though they often make tools from branches of the same trees in which they hunt galagos. They use two main tree species, one of which is also the most common tree in their home range, *Pterocarpus erinaceus*, or kieno, in the Malinke language. The other is not as common; based on its relatively scarcity, it is apparent that the chimps go out of their way to find a *Terminalia*, or wolo (in Malinke), tree in order to modify a live branch of this species for a spear. The chimp may leave the tree of interest, find the nearest wolo, and return to the tree of interest to make a spear and hunt with it.

Fongoli chimps are consistent in their tool making, at least once they get to a certain age. The lengths of the 335 tools we have retrieved and measured provides one example. Adult and adolescent males and females at Fongoli make similar-size tools, averaging about 70 centimeters in length, but youngsters do not begin to make adult-size tools until they are about four years old. Chimp babies make tiny, undersize tools that are proportionate to their hands, averaging 37 centimeters in length, and they use these to seemingly hunt in earnest, even though such tiny tools (literally, not much bigger than twigs in the case of babies younger than two years) have no utility when it comes to bush-baby hunting.

When young chimps reach the age of four, they begin to make longer tools, but the sizes of their tools still really varies. This is around the time that infants at Fongoli are weaned. Juveniles' tools average almost 50 centimeters in length, but range from less than 30 to more than 80 centimeters long. Some are too long, catching on the tree cavity and breaking, being too unwieldy for juveniles to use. Young chimps may even make a spear from a vine, which is too flexible to make a good hunting tool. At this age, chimps are kind of all over the place when it comes to fashioning these tools, indicating to me that they're using trial and error to ultimately come to a tool length that will give them a successful hunt.

Of course, the depth of the cavity and how far down the bush baby is should have some influence on the characteristics of the tools made, and this is another aspect of Fongoli chimpanzee hunting that we've investigated too. One of the Fongoli researchers, Jacques Keita, is a very good climber, and he has scaled numerous trees to take measurements of the depth of bush-baby cavities, as well as recording if these cavities did house galagos. Jacques worked closely with me on our galago-cavity study, and I distinctly remember talking to him one day about where I had placed a temperature logger in a kieno-tree cavity. I told him I had placed another logger in an adjacent tree, but by the time I was noting it was probably a good idea that he climb down from the tree he was in and climb the neighboring one, he had already climbed through the canopy to the second logger. For a year, Jacques checked a sample of cavities that chimps had hunted in. He used a borescope, which is an instrument that plumbers use to examine pipes, with a light and a camera at the end of a flexible wire. Jacques also measured the depth of water in these cavities to see how much or if they became inundated after rains, which is also when we see an increase in chimpanzees' tool-assisted hunting.

Even though I have a hard time climbing to explore the cavities that the Fongoli chimps hunt for galagos in, these cavities are not very high, in general. This mirrors the low height of trees at Fongoli—around 10 meters on average—compared to forests where apes are studied, where the tree canopy can be up to 50 meters tall. Of the 148 cavities we've measured, they were five and a half meters high on average, ranging from one and a half to 20 meters off the ground. Trees that the cavities are found in average 10 meters tall. I once hauled myself up to a 1.5-meter-high cavity after about five minutes of struggling, only to find that there was no evidence of a galago inside. As I jumped (okay, fell) back down, I saw that alpha male David was watching me intently. Sometimes, I imagine that this was one of the reasons David periodically charged me to see if I would submit—run away and scream—like all the other chimps. Maybe he wondered, "If that thing can't even climb, surely I really can't be afraid of it."

I also censused trees in the Fongoli chimpanzees' home range to estimate the density or availability of potential galago-nesting sites here. At Fongoli, we use a transect to record phenology of food plants that the chimpanzees feed on—and this is basically a line just over 3 kilometers long and 20 meters wide that cuts through the chimpanzees' core range. A researcher walks this line each month and records the reproductive state of seven hundred–plus large trees (trees that have a trunk size of at least 10 centimeters in diameter) in this area according to whether they have fruits, flowers, and leaves. This gives us an estimate of the foods available to the Fongoli chimpanzees across months and years. At any rate, I used this transect as well as chimpanzee trails to estimate the number of potential galago-nesting cavities. I surveyed sixty-three different trees belonging to nineteen different species, and found that only 19 percent of them had branches or trunks with cavities that looked suitable for bush babies to nest in, based on estimates from my observations of hundreds of chimpanzee tool-assisted hunts. It's likely that my view of these potential cavities from the ground overestimated the number that were in fact suitable for galago nesting, so the actual number of potential bush-baby nesting cavities was surely lower. This suggests that chimps at Fongoli do not simply run into galago nests by accident or with great ease; some degree of searching is required in order to hunt these tiny primates. In fact, we've recorded a number of trees and particular cavities in them that are targeted year after year. One of these trees (which I call "Lucille's tree" because adult female Lucille was the first chimp I saw hunting there) was such a reliable hunting spot that we would hurry over to it to see if we could catch someone hunting when the chimps traveled over the small hill it was on to get to the Kerouani stream. Sadly, Lucille's tree was cut down in 2019—large-scale timber harvesting has targeted larger trees in the Fongoli area, stemming from the increased human population in the Kédougou region that accompanied the recent gold rush.

SEASONAL PATTERNS

In looking across the five-hundred-plus tool-assisted hunting cases from Fongoli, it is apparent that this behavior is seasonal, taking place mainly during the early rainy season. I have shifted my ideas about why this is over the years. In the first paper we wrote about the behavior, I hypothesized that this seasonal effect might be due to increased competition between males and females, since this is also a time of year when the whole community can be found ranging together. Such increased feeding competition might lead females to broaden their feeding niche. However, over the years I have become more and more convinced that the behavior of the galagos themselves is likely responsible for this seasonal effect. The rainy season at Fongoli is characterized by big storms, especially at the beginning of the season. Rain falls in such torrents that tree and branch cavities become inundated with water. Fongoli chimpanzees often hunt following a storm, and I believe that the galagos cannot withdraw into the recesses of their woody cavities to avoid chimpanzee hunters when these cavities are filled with rainwater. Jacques found that the galago cavities he monitored indeed usually had water in them following rains. Dry-season hunting is so rare in some months that I also think galagos must resort to using the depths of tree-trunk or branch cavities or perhaps more frequently utilizing only tree-trunk cavities, which are less accessible to chimpanzee hunters. The high temperatures during the Senegal dry season exceed 46°C (115°F), and the exposed branch cavities especially may become too hot to be bearable by these diminutive primates. In order to test this hypothesis, I have placed small temperature loggers in tree-branch and tree-trunk hollows and measured temperatures across the seasons at Fongoli, which we will examine to see if there are significant seasonal differences that would affect whether and how galagos use certain types of nesting cavities.

Alternatively, there may be a nutrition-based explanation for the highly seasonal aspect of galago hunting at Fongoli. At other chimpanzee sites, like Kanyawara in Kibale National Park in Uganda, chimpanzees hunt most when there is more ripe fruit available. Galago hunting at Fongoli also takes place alongside the availability of one of the chimps' top fruits, *Saba senegalensis*. The explanation from Kibale is that the costly monkey hunting seen there is made possible by a high-fruit and thus high-energy diet. I doubt the same can be said for the relatively low-cost galago-hunting behavior at Fongoli.

· 5 ·
Risks on the Savanna

SNAKES, BEES, AND
HIPPOS, OH MY!

The Fongoli Savanna Chimpanzee Project protocol includes a list of risks, which I give to researchers, students, or other visitors who plan to come out to my research site. Many of the risks are in the form of diseases common to the area, like malaria or dengue fever, or stresses associated with heat and dehydration related to the hot and dry climate in southeastern Senegal. Many of the dangers associated with fieldwork at Fongoli are also those that the chimps experience.

Chimpanzees in a savanna landscape are exposed to some of the same risks faced by apes in forests—leopards as predators, for example, or aggression from neighboring chimpanzee groups or even from individuals within their own social group. But other risks at Fongoli are specific to a savanna environment. These range from hazards like wildfires to frequent encounters with multiple species of dangerous snakes and potential predators like hyenas.

One of my fondest memories is of the time adult male Bo seemed to warn me about getting too close to a snake at Sakoto ravine. He would shake a branch whenever I approached a tree with a snake in it that I was trying to identify. I was pretty sure the snake was nonvenomous, but I wanted to get a better look at it. Bo seemed to warn me twice. When I continued to get closer to the snake, he finally jumped out of the tree he was in and ran off, looking back at me to follow, as chimps do to one another. I actually felt bad about not heeding his warning. I wondered if Bo was questioning how humans have lasted so long on this planet.

POTENTIAL PREDATORS AND OTHERWISE
SCARY ANIMALS

Even though potential predators like lions and wild dogs have been extirpated at Fongoli, others like leopards, spotted hyenas and rock pythons are still risks that the chimps face here—as do some humans, on a different level. There are also other animals that, at least to a human observer, do not seem to pose a threat to chimps, but that the Fongoli apes treat as a valid danger. These include domestic dogs, common jackals, hippos, and even tortoises and turtles. If we think hard about why the chimps might interpret these animals as risks, though, there is usually an underlying explanation. Encounters between primates and potential predators are notoriously difficult to observe, given that the presence of a human usually serves as a deterrent to that predator, which is not habituated to us like the chimps are. However, we have been able to record some of these encounters at Fongoli.

Senegalese researcher Waly Camara, Mbouly's son, who worked for the Fongoli Project for several years, witnessed the Fongoli chimps finding a leopard in a small cave one day, on a rocky slope at a place we call Djendji Plateau. He watched this encounter for just over an hour. Leopards are the major predator of many primates across Africa and Asia, including chimpanzees. Waly had first heard the chimpanzees screaming near a small cave and found two older chimps, adult male Bandit and adult female Farafa, using long, heavy branches to thrust into a small opening among the boulders. Bandit was at least thirty years old at the time, and Farafa was about forty years old. She had an infant, Vivienne (who was two years old), clinging to her belly throughout the ordeal. Bandit and Farafa alternated their place at the mouth of the cave, stabbing the dead branches in over and over. As Farafa was doing this, Waly heard a loud roar from the cave and thought it was a lion. All the chimps fled up into the surrounding trees, while Waly was left trying to hide somewhere in case whatever it was came his way. About forty seconds later, a leopard fled from the cave, heading in the opposite direction of the chimps, and Farafa, adult son Mamadou, and adult male K.L. chased after it. They returned shortly after, and all the chimps proceeded down the slope to drink at the Djendji water hole. Waly was able to collect one of the tools that was used against the leopard, and it was much bigger than the spears made to hunt galagos. While galago-hunting tools average about 70 centimeters in length, the leopard-stabbing tool was almost twice as long and much thicker than the average galago-hunting tool. It was also much heavier.

I believe Bandit and Farafa were the ones at the forefront of the leopard mob because they likely had more experience with these big cats, given their advanced age. Elders in Djendji village told me that some decades ago, people were unlikely to walk the 3 kilometers between their village and the neighboring one without hearing the typical sawing-wood call of a leopard. While we no longer hear these calls at Fongoli, at least one female leopard continues to make her territory here, and the chimps treat her as a threat. Camera traps set up at Sakoto pool recorded a young adult female leopard frequenting the pool area over the course of several days, moving down and probably into the cave, during which time the Fongoli chimpanzee community moved to the far edge of their range, approximately 8 kilometers east. A visiting researcher, Dr. Nicole Herzog, saw this female leopard as she crouched in a small tree, watching green monkeys foraging in a recently burned area near Sakoto ravine.

In savanna ecosystems, chimpanzees face additional predators, such as spotted hyenas, that are not usually encountered by forest-dwelling apes. We have seen Fongoli chimps interact with spotted hyenas several times and have heard them react to the calls of hyenas at night by giving a chorus of wraaa-barks or alarm calls. Spotted hyenas are roughly the size of the largest domestic dog, weighing up to 165 pounds, and their jaws are literally used for crushing bone. While hyenas pose a relatively small threat to humans, they are known to attack people who are incapacitated in some way, including sleeping. Reports of such events can be heard in this region of Senegal, and I am certain I was once stalked by at least a curious hyena one night when I left the chimps at their nesting site after dark and got lost in a sea of dry elephant grass on my way home. We had heard a hyena that morning in this area, and, as I abruptly stopped in the grass, I heard something big with a four-footed gait stop instantly. I stopped flailing through the grass, gave a chimpanzee wraaa-bark, and started for home in a less erratic way—but it still makes the hair stand up on my arms when I think about that moment. I cannot think of any other animal it could have been.

Twice, we have witnessed the Fongoli chimpanzees aggressively chasing hyenas and throwing stones at them, while another time, a lone adult male ignored a spotted hyena that was nearby. Two of the encounters were witnessed by Dr. Kelly Boyer Ontl, who served as Fongoli Project manager in 2008. Kelly and Michel Sadiakho were following adult male Bilbo through a woodland area, when an adult hyena walked and trotted past them only about 30 meters away. Bilbo turned and stood up on two feet to look at the hyena, and just watched as it passed. Another time, it was right after dawn when the chimpanzees were waking up and coming down from their nests. Three adult males were in a subgroup that day, and they had recently woken up upon hearing pant-hoots from another subgroup of chimps who had nested nearby. The three males immediately ran in that direction, Kelly and Michel following. When Kelly and Michel arrived five minutes later (the chimps had gotten to the scene much quicker), they saw three hyenas running across the open grasslands away from the chimpanzees. The entire chimp group gave warning barks or vocalizations, and one adult male, his hair standing on end, ran after the hyenas on two feet. The other chimpanzees followed. The group continued to call and display while a second adult male also ran bipedally after the hyenas and threw two large rocks in their direction. Using projectiles as weapons against potential predators, baboons, and even other chimpanzees, is not uncommon at Fongoli—not surprising, given how many stones litter the landscape here.

I was with a large subgroup of more than a dozen chimps at the Tukantaba ravine in another encounter where they threw stones at a hyena. I heard Luthor, a young adolescent male at the time, giving an alarm call I didn't recognize. The chimps, like many other primates, have specific alarm calls for different types of predators. All the other chimps stopped what they were doing and looked in one direction from where, a few moments later, a bushbuck antelope came running as fast as I've seen anything run! I briefly thought that it was just like Luthor to alarm call at a harmless bushbuck——he would also alarm call at my flashlight if I didn't remember to hide it, and, once, he alarm called at the reflection from my water bottle. But an instant later, a spotted hyena flew through, hot on the heels of the bushbuck. The chimps charged the hyena, throwing stones and hooting. I distinctly remember Diouf, the largest adult male chimp at Fongoli, standing up on two feet and throwing stones as the hyena came within about 15 meters of where we all stood. That hyena literally turned tail and ran back the way it came, with a number of the chimps chasing it.

I tried to keep up with these chimps to record what happened, even though the idea that the chimps could catch the hyena was pretty laughable. I remember looking to my right and seeing young adult female Lily running as if she were genuinely going to get that hyena. At about that moment, we flushed a female patas monkey out of the bushes. Lily started after the patas monkey instead, which was just as laughable—patas monkeys are the fastest primate, even though a few Fongoli chimps have been able to catch them over the years, when the monkeys get trapped within tree crowns. It was like some bizarre scene from a Dr. Doolittle movie, where different animals kept popping out of the scenery. The chimps gave up the hyena (and patas monkey) chase pretty quickly—but, upon returning to the rest of the party, I found that adult female Nickel had captured a bush baby! I could hardly keep up with writing all of this down.

It's interesting that the chimps have reacted aggressively toward spotted hyenas, which are many times larger than the domestic dogs they also have to deal with. But many different primate species "mob" the predators larger than they are. The more interesting aspect of this behavior is that the dogs are not mobbed in the same way as larger and more dangerous potential predators. In fact, the common jackal found in this area of Senegal is usually ignored by chimpanzees, sometimes chased and even appearing to annoy some of the chimps with their presence. I once witnessed Diouf throw a rock at a young jackal that approached him curiously as he was resting—similar to how I've seen David and other chimps treat warthogs that approach them too close. But former alpha male Foudouko once spent around thirty minutes warning barking at an adult jackal playing by itself out on the plateau at Sakoto. David even chased away a banded mongoose that was cracking open snail shells on a rock not too far from where David was resting once, so sometimes chimpanzees just appear to get annoyed by other critters. Similarly, I've seen K.L. get up and chase away a noisy turaco bird from a tree he was trying to rest in.

The village dogs in this area of southeastern Senegal are roughly the same size as jackals. Chimps are usually fearful of domestic dogs—roughly the size of a beagle or other midsize domestic dog—when they first encounter them. This seems to be related to the fact that dogs are often with humans and sometimes involved in a hunt guided by humans. We know that adult female Tia was attacked by dogs when her infant Aimee was taken by young men who had been hunting warthogs with the dogs (as I describe in chapter 6). People often hunt green monkeys and patas monkeys with dogs as well. We've been around the Fongoli chimps numerous times when they have encountered domestic dogs. If the dog is not with humans, or at least some distance from humans, the male chimps in particular may then regroup and chase it after initially fleeing. Alpha male David was especially good at this, and Dawson is another chimp who seems to go out of his way to charge dogs if they foolishly continue to bark at the chimps—for example, when they feed at baobab trees near people's seasonal fields.

I haven't seen the chimps harm a dog yet, but I did hear one yelp one day when David kept charging it when we were at Tounou ravine. Maybe he grabbed it, or perhaps he threw something at it. The woman who owned the dog—she had been doing laundry at a stream where the chimps were—finally called it away.

A recent case involving adult male Lex would have probably had a dire outcome if Michel Sadiakho and others hadn't been near him on that day in March 2023. I had just left Fongoli after a brief visit with a documentary film crew who wanted to interview me in the field. Michel and members of a film crew were with a relatively small party of several adult males and a couple of adult females and their infants. They heard screams from Lex and found him literally covered in these small dogs. There were so many of the dogs that Lex had been overwhelmed and was prostrate on the ground. Michel and the others chased the dogs away, and Lex ran away as well. Michel was afraid that Lex had been too injured to survive, but they found him again a few days later. His wounds seemed superficial and mainly on his hands and feet (as far as they could see), but Michel is firmly convinced that Lex would have suffered a horrific end if the research team had not been there to intervene.

While hippopotamuses do not attack chimpanzees, they are treated as a danger by Fongoli chimpanzees in the rare event that they are encountered. Hippos are very large, herbivorous, aquatic mammals, with fully grown adult females weighing about 1,300 kilograms and fully grown adult males weighing up to 4,500 kilograms. Hippos are actually one of the most dangerous animals in Africa, killing more people per year than most others, but these lethal encounters usually take place when humans are in canoes or small boats in the hippo's territory. These animals can be especially dangerous on land at night too, when they leave the water to feed. I believe that part of the reason that the chimps react so strongly to hippos is that there are so few left in the part of the Gambia River that

Figure 5.1. Adult male Lex stands on the edge of the Gambia River at the area we call Dafoula, where the Fongoli chimpanzees have encountered hippos. Photo by Katie Gerstner.

forms one edge of the northeastern border of the Fongoli community range.

For more than a decade, we only saw hippo tracks emerging from the River Gambia and did not witness encounters between Fongoli chimps and hippos, but in 2019, I was with a large party of chimps at the Gambia River, at a site we call Dafoula, and at least one hippo was in the river there. A large, unfamiliar "water thing" seems to evoke incredibly strong feelings from the Fongoli chimpanzees. This was perhaps one of the most intense reactions I have seen from the chimps to a threat. At the sound of the hippo surfacing and blowing air out of its nose, the entire chimpanzee party screamed like I have never heard them scream! They frantically reassured one another and perched atop the small shrubby trees bordering the river, staring intently at the place where the hippo then submerged. Around about the time the chimps settled down, the hippo would come up for air and the cacophony began again. Hippos can hold their breath for quite a long time but need to come up for air at least every five minutes. I actually felt a little sorry for the hippo.

The next time I was with chimpanzees that passed this same site, they sat for some time along the edge of the river here, scanning, but I didn't see or hear the hippo again, and they didn't indicate that they had either. In June 2021, Michel was with the chimps at the same spot when they had an even more memorable encounter with a hippo, because this particular hippo would come up for air and look at the chimps. The same display of screaming and warning barking at the hippo ensued and again lasted for hours.

It may be that the Fongoli chimps will habituate to the presence of hippos here, if they encounter them often enough. Given that hippos have used relatively nearby areas of the Gambia River for years, though, it's surprising that at least the older chimps aren't less reactive to them. As chimpanzees become more familiar with threats in their environment, they usually become less reactive—but that also depends on the nature of the danger. Leopards will likely always provoke an extreme response, while other dangers are a type of risk that means something entirely different for the chimps.

BEES

Of all the animal threats, I should mention one of the most dangerous: bees. Each year in the Fongoli area, we hear of a person who has been fatally attacked by bees, eventually succumbing to hundreds of stings. It is not something most people in the United States probably worry about, for example, unless they are allergic to beestings. In southeastern Senegal, however, fatal bee attacks are a reality, as is lethal snakebite. Bees are very common in this area, and given how many animals—including humans—raid their hives, they are very aggressive and easily roused. They are also even more likely to attack during the dry season, when it is especially hot. You cannot walk more than a mile without passing a natural beehive in this part of Senegal, and if you are unlucky enough to encounter one that has been recently disturbed, you are in a very dangerous position.

Although Fongoli chimpanzees are averse to beestings as well, they risk a lot for honey. That's not surprising, given how nutritious—and tasty—honey is. During the dry-season months especially, they go out of their way to raid beehives. About 65 percent of beehive raiding for honey takes place at this time of year, and, as a researcher, you quickly learn

Plate 1. Foudouko was the first alpha male identified via pant-grunt vocalizations in the Fongoli community, around the year 2005. He was overthrown as alpha during a fight involving all adult males in late 2007. He and his close coalitionary partner, Mamadou, were wounded in this fight. Foudouko was subsequently ostracized from the Fongoli community by other males for several years and was initially assumed dead by researchers. After trying to reintegrate into the community some years later, he was apparently killed in a concerted attack by Fongoli community males in 2015. He was alpha for at least two years but likely longer. Photo taken in 2007 by Frans Lanting, © Frans Lanting | Lanting.com.

Plate 2. Yopogon inherited alpha status in early 2008 after Foudouko was ousted in the fight involving all adult males. He disappeared in 2010 but was alpha for about a year and a half. Photo taken in 2007 by Frans Lanting, © Frans Lanting | Lanting.com.

Plate 3. Lupin took over the alpha position from Yopogon in 2009 and lost it to David in 2012. He was alpha for two and a half years. Photo taken in 2013 by Dondo Kante, courtesy of the Fongoli Savanna Chimpanzee Project.

Plate 4. David (being groomed by younger adult male Bo) took over the alpha position from Lupin in 2012. He lost alpha status in 2018 after a concerted attack, likely by most adult males in the Fongoli community. He died from his wounds ten days later. He was alpha for six years. Photo taken in 2013 by the author.

Plate 5. Jumkin inherited alpha status after David died in 2018. He lost his alpha rank to Cy in 2022. He was alpha for four years. Photo taken in 2023 by Carson Black.

Plate 6. Cy took over the alpha position from Jumkin at the very young age of 14. Cy was still alpha as of January 2024. Photo taken in 2019 by McKensey Miller.

Plate 7. Head Fongoli Savanna Chimpanzee Project researcher Michel Sadiakho observes chimpanzees during the wet season. Photo by Nicole Wackerly.

Plate 8. Head Fongoli Savanna Chimpanzee Project researcher Michel Sadiakho follows chimpanzees across a grassland during the dry season. Photo by the author.

that if you see multiple chimps running in a relatively helter-skelter way, you better run too. If you are stung by a bee, it releases a chemical signal that its hive mates use to locate it and continue the attack. The situation quickly escalates when a beehive has been raided and hundreds of bees are retaliating. At the beginning of the Fongoli Project, when this happened, I would immediately worry that I had somehow frightened the chimps and this is why they were running. My inaccurate assumptions cost me several seconds of escape time but, after multiple cases of getting many stings, I started retraining my reactions, and it has enabled me to get a better look at the chimps' beehive-raiding and honey-eating behavior.

As with almost any behavior, there are individual differences among the Fongoli chimps in beehive raiding. In two hundred cases, out of a sample of almost three hundred beehive raids for honey, we were able to identify the chimps who raided the hives. Young adult male Dawson leads the Fongoli chimps in beehive-raiding activity, with almost twice as many as the next most prolific beehive raider, older adult male Bilbo, who disappeared in 2019. The female with the most beehive raids is Tumbo, who is currently alpha female, tying for sixth place in successful beehive raiding with oldest and lowest-ranking adult male Siberut. Given the stings these chimps endure, honey must be considered a pretty high-quality food. However, even for those individuals who don't raid beehives themselves, there's a chance they can get access to honey via sharing or scrounging.

In some cases, we have seen individuals who raided beehives sharing the honey. In other cases, chimps scrounged small pieces of honeycomb from the site where honey was eaten by other chimps. They will pick up what look like even tiny pieces from the ground where someone has been sitting and eating honey; in addition, they'll lick dripped honey off leaves. Even though bits of honey and comb may have fallen to the ground, if the original "owner" is sitting on the ground at the site, other chimps will hesitate to retrieve it if it seems as if the owner is still interested. (We see this in the case of meat scraps too.)

Even though the chimps at Fongoli seem to share plant foods more than at other sites where chimps have been studied, foods they most often share are high-quality, hard-to-acquire, and preferred ones like honey, meat, and baobab fruit. At most other study sites, food sharing among unrelated chimps is usually just meat. We have not yet quantified the amount of honey available to chimpanzees at Fongoli, but the frequency with which they encounter hives suggests it may be more commonly encountered than in forests where chimpanzees live.

And, when we compare our observations to those at other sites, the Fongoli chimps are rarely deterred from raiding beehives, only turning back if the hive has been previously disturbed by another chimp and the bees are attacking anything that moves nearby. Even then, certain individuals (like young adult male Lex) will take the brunt of what must be multiple stings from bees that make it through the chimp's long hair, to escape with a handful of honeycomb. The fact that I've heard many of the Fongoli chimps scream when they are stung by bees also indicates to me that beestings are not taken lightly by the chimps either.

Over the years, I have become more sensitive to beestings, so I carry an EpiPen with me. One of the worst times I've gotten stung is when at least forty bees stung me one day at Point D'eau, the small spring at the base of a flat-topped hill that is the chimps' permanent water source. I was stung after a chimp raided a beehive down near the water hole—I made the mistake of fleeing in the same direction as the chimps, *up* the ravine's slope, which is about a hundred meters of incline that often requires using your hands to climb. About midway up, I encountered *another* angry hive that chimps had just raided. By the time I crawled onto the edge of the plateau, I had so many beestings that my throat started to swell shut. I had to keep running for a while, as the bees hadn't given up yet, and this was one of the first times I took notice of adult male Bandit's strategy. As I pulled myself up onto the plateau and tried to run further, I saw Bandit just about 5 meters from me. He startled, and I think he would have dashed off, but he immediately slowed down and walked in the opposite direction. However, his hair was standing up, as it does when chimps are excited in some way—scared or angry, for example. It seemed like he purposely slowed down so as not to attract more bees his way, even though his first inclination was to move more quickly out of my way.

I've seen Bandit do this numerous times now, and I've tried to mimic his behavior if I can. The key is to be just far enough away from the bees so that you're not one of the things that they are mobbing—and it helps if there are other running chimps (or humans) for the bees to track onto. If you're within 50 meters of a disturbed hive, you need to run no matter what. If you're at least 75 or 100 meters away, you might walk away without getting stung if there are other running chimps/humans around to distract the bees.

One day, I was unfortunately getting into a brushy thicket when I ran into a disturbed hive. It was hard for me to run through the thick scrub, and I kind of fell out headfirst over the edge of a rocky hill. I remember seeing adult male K.L. sitting a short distance off—apparently out of range of the bee storm—eating leaves and watching me like a moviegoer with their popcorn. I'm pretty sure most of the chimps have seen the bizarre way I run away and hit at bees. Maybe this is why a couple of them seem to go out of their way to make sure I am farther away from them right before they raid a beehive.

One of my favorite chimps, Bilbo, really endeared himself to me one day as he traplined over half a dozen potential beehive sites. I was following him when he exited his night nest in the morning until he made another night nest at the end of the day. Bilbo broke off from the rest of the subgroup that day and explored at least seven different sites where there had been, or possibly could be, honey. He ended up raiding two hives. At the first one, as he stood on the tree branch about to thrust his arm into the cavity and grab a big handful of honeycomb, he looked pointedly at me. Chimps rarely do this. They usually ignore us or even pointedly ignore us, especially adults. But Bilbo kept looking at me. It wasn't a threat, and I was kind of clueless, and then I realized he was waiting for me to do something. I moved further away from the tree and, once I did, he crammed his hand in and grabbed a bunch of honey. Then we both hightailed it out of there. Apparently, his judgment was better than mine because I didn't get stung but probably would have if I would have stayed closer to the tree.

Another time, I was with a graduate student and adult male Lupin looked at us in this same way. We were in the middle of an old field where bushy vegetation had started regrowing, and Lupin was standing on an old tree trunk that had fallen over. I really didn't think that there would be a beehive there, so I didn't catch on at first, but I remarked to the student that Lupin was looking so oddly at us. Then he raided a beehive in the fallen tree trunk, and we had to run like hell. I appreciate Lupin's consideration, although he wasn't as patient as Bilbo and gave us only a few seconds to figure out what he was about to do. Humans are so clueless sometimes. It's a wonder we are doing so well—at least, in terms of how many of us there are compared to other great apes.

Even though we have seen Fongoli chimpanzees raid hundreds of beehives, it was not until almost twenty years after our study began that we saw a chimp take advantage of a bird called a greater honeyguide (*Indicator indicator*). Honeyguides are birds that have been described as engaging in mutualistic relationships with people in some parts of Africa. Anecdotally, there have been reports that other animals, like honey badgers, also follow honeyguides to beehives, but more extensive investigation has revealed good evidence that only humans use these birds in a commensal relationship. People in East Africa who follow honeyguides to beehives traditionally leave a large piece of honeycomb (which includes the larvae the birds eat) stuck onto a tree or shrub after collecting honey. I have also been solicited by honeyguides—and have followed one for some distance—until it led us to a beehive. I felt remiss that I couldn't raid the hive.

I've seen these birds try to solicit chimps to follow them at least a dozen times now. In fact, adult male Bandit even tried to swat away a honeyguide that was flying around and soliciting him at a termite mound where he was termite fishing, chattering at Bandit with the typical guiding call, moving out and back in the same manner they use to solicit humans: the honeyguide continues its calling and flies out and back, moving ahead of you as you follow it, ultimately to a beehive. In late 2020, Michel was following adult male Dawson as his focal subject when he saw Dawson follow a honeyguide to a bees' nest, raid the hive, and escape with a big handful of honeycomb. Dawson did not otherwise interact with the honeyguide, but it's possible the bird was able to scrounge honeycomb or larval remains after Dawson's beehive raid, much as other chimps do when someone is eating honey in the group.

DANGEROUS REPTILES

One thing that strikes me when I see the Fongoli chimpanzees find a dangerous snake is that they seem so human in their need to get a good look it. Almost every individual in the group has to come check it out; some stay and harass it, while others seem satisfied to simply look at it briefly and leave. But there seems to be a need to *see* it. I noticed the same phenomenon when I studied patas monkeys in Kenya for my dissertation research in the early 1990s. Once, the entire group of thirty-eight monkeys circled around a puff adder (a type of viper we also have in Senegal), and many of the monkeys were standing on two feet, peering at the snake. I saw a juvenile join the circle; it must have bumped some other monkey, because it set off a chain reaction of heebie-jeebies that had all the monkeys literally jumping in the air.

At any rate, there are numerous studies of nonhuman primates and their reactions to deadly snakes, many of them focusing on the degree to which fear of snakes is learned or ingrained. We have also done research on the subject at Fongoli because the chimps seem to encounter dangerous snakes at higher rates than has been reported elsewhere for apes. Humans in this area, too, are subjected to a relatively high rate of mortality due to venomous snakes, with almost one-third of accidental deaths caused by snakebite. (Worldwide, according to the World Health Organization, only a fraction of 1 percent of total human mortality was due to snakebite between 1990 and 2019.)

So far, we have observed more than fifty encounters between Fongoli chimpanzees and dangerous reptiles. When you take into account how many individual chimps have ultimately interacted with these reptiles, we have more than 175 cases where we can look at how each chimp reacted to them. Chimpanzees at Fongoli regularly encounter puff adders and rock pythons, but they also run across cobras and black mambas fairly frequently. Less often, they encounter (or at least react to) smaller vipers like the deadly saw-scaled viper and the night adder or the relatively harmless ball python (which the chimps nonetheless reacted to very strongly!). We have seen them encounter tortoises periodically too; even though these reptiles do not seem to pose a threat, the chimps treat them similarly to the dangerous reptiles they encounter. They also seem to treat reptiles according to the setting. For example, if a reptile that usually elicits little reaction, like a monitor lizard, is in the water, the chimps react much more strongly. And I once saw adult female Natasha chase a monitor lizard out of a tree crown that an adult male chimp had just chased her into. I'm not sure if this was redirection on Natasha's part or if she was genuinely afraid or just annoyed by it.

A single turtle has effectively kept all the adult males from soaking in Sakoto pool—as discussed earlier, one of their favorite pastimes during the early part of the rainy season, when the heat and humidity is still really high. When adult male Lupin was alpha male, I remember him exhibiting distinctly *un*-alpha-like behavior because of that turtle. He literally hung from a vine above the pool, staring at the offending turtle and then dipping a toe in before gathering the courage to get in the pool alongside Mamadou, who was always one of the adult males that really seemed to like to soak there. That same day, Mamadou was the first male to brave the turtle-infested waters of Sakoto, and he sat at the other side of the 3-meter-wide pool, staring down into the water and finally smacking his hand down hard on the surface of the water above the turtle.

I imagine that this relatively greater fear of harmless reptiles in water stems from the fact that crocodiles are found in the Gambia River, which borders one edge of the chimps' range, as well as in its tributaries where they drink, like Fongoli stream. In support of this hypothesis, almost all the chimps drink from the Gambia River while hanging off a tree or vine, above the water's surface. They do not get into the river to cool off and drink, as they do at Sakoto pool and all the smaller streams within their home range.

The exception is Cy. Cy was born in 2008 to Tumbo, and he was her first offspring. Tumbo didn't have a second infant (Zoey) until almost a decade later—an interval between successive births that is atypically long for Fongoli chimpanzees. This meant that Cy received his mother's undivided attention for many years. Cy has always exhibited behavior that seemed to leave his very tolerant mother exasperated. One such behavior was his tendency as a juvenile to hang from a vine into the Gambia River and soak. No other chimp has been observed to do this, and Tumbo would sit nearby, seemingly agitated and unable to lure Cy into following her away from his cooling-off spot. Fortunately for Cy, the crocodiles who are left in this part of the Gambia River are too small to consume even a juvenile chimpanzee, but this doesn't mean one might not have tried. However, Cy has survived to young adulthood and moved on to exhibit other interesting, if not atypical, behaviors.

One of the most interesting aspects of the Fongoli chimpanzees' interactions with potentially deadly reptiles is that there are distinct individual differences in the chimps' reactions. In fact, if these chimps were people, we would likely consider some of them to be experiencing ophidiophobia (having a phobia or fear of snakes)! Out of all the Fongoli chimpanzees we have seen interact with dangerous reptiles, adult males Dawson and Jumkin (former alpha) react most strongly. Dawson, without fail, uses stones and sticks to attack reptiles, while Jumkin often does as well. But, as an adolescent male, Cy is the only individual we have seen so far that has killed a reptile. Dr. Landing Badji, the assistant director of the Fongoli Project, observed Cy using stones as projectiles to kill a small viper, which turned out to be the very deadly saw-scaled viper. And we have recorded at least one case of chimpanzee mortality due to snakebite, which I tell about in the next chapter—after the story of infant Aimee's rescue from poachers.

· 6 ·
Neighbor Apes

CHIMPANZEES IN A
HUMAN LANDSCAPE

On January 24, 2009, I was in the US when the manager of the Fongoli Project, Dondo Kante, called me from Kédougou. He reported that a young man had approached him, asking if he knew anyone who would be willing to buy an infant chimpanzee. This young man—a high school student in Kédougou—had been out hunting warthogs when he and a friend and their hunting dogs ran across some chimpanzees. At this point, Dondo told me, he began to explain the goals of our research project, but he realized the young man became nervous, so he just asked to see the baby chimp.

Dondo was appalled to see how many people were coming by to look at the young chimpanzee, who was in a dark room in the town market. He eventually pieced together what had happened from the information he got from the hunters, including when he visited the capture site with one hunter. I have no doubt that we would have never received all the information if a foreigner instead of Dondo had first been approached by the hunters.

Dondo learned that the hunters' dogs had attacked the baby's mother beneath some *Saba senegalensis* vines. We believe it was at this time that the infant fell from her mother or was grabbed by the dogs, as she had shallow puncture wounds on her back when Dondo took her from the hunters. We believe the mother would have escaped from the dogs initially by climbing a tree, but was probably frightened as the hunters arrived. We've been with the chimpanzees when they encounter dogs; males are able to chase the dogs away, especially when they are in a group with other males, although they initially respond by climbing trees

Figure 6.1. Author (*right*) and Fongoli Project manager Dondo "Johnny" Kante, man of all trades and savior of Aimee. Photo by William Aguado.

to escape from them. We think the mother was treed by the dogs but then leaped from the tree to escape the humans. At this point, the hunters said, their dogs attacked her, and her infant was torn off or fell off during this attack. The hunter noted that the mother turned and attacked the dogs, chasing one of them from the scene, but was again pursued by another dog as she ultimately fled the area.

Based on the description of where the men had gotten the chimpanzee, we concluded that she was in all likelihood from the Fongoli chimpanzee study group. I suggested we try to return the infant to her mother, if she could be located. Otherwise, we should determine if the baby was old enough to possibly survive on its own. Chimpanzees have been known to survive without a mother if they are at least two years old, although this is rare, as they usually aren't weaned until they are at least four years old.

We agreed that Dondo would try and get the infant from the men. He led the hunter to believe that he might find someone interested in buying the baby. In trying to understand the circumstances of the infant's capture, Dondo kept in communication with one of the hunters in the days following. At first we had wondered (hoped?) that the infant might be older infant Teva, Nickel's first daughter, who would not have been nursing too frequently. But then, we saw adult female Tia, who was injured and without an infant. We ultimately confirmed that the baby was her infant, Aimee.

Aimee was about nine months old at the time she was captured, having been born in May 2008. She was the daughter of first-time mother Tia. An infant of Aimee's age would be nursing hourly, so we were also concerned about her nutritional state. Her captors had tried to give her some banana, but she was afraid to take it from them. Happily, we were later able to get her to take food.

The day after Dondo first called me, I received photos of the baby and confirmation that he had her in his care. (We didn't compensate the hunters for the infant chimpanzee, of course, which could have encouraged active hunting of chimpanzees to acquire babies for the pet trade.) She had abrasions over her left eye and matter in the same eye that made it impossible for her to open it. She was kept in a hut at Dondo's house in a large cage he borrowed from a man who sold birds. Dondo fed her milk that he prepared from regular milk powder whenever she cried; otherwise, she was isolated from people so as to minimize the likelihood that she would become ill.

After learning of the baby chimp's capture, I contacted Janis Carter of the Chimpanzee Rehabilitation Project in The Gambia to discuss the situation. Janis had successfully reintroduced an infant chimpanzee into a wild, unhabituated group in the area between Kédougou and Salemata in southeastern Senegal. She had come to Africa more than three decades before to bring Lucy, a captive chimpanzee who had been taught American Sign Language in the United States, to a sanctuary in The Gambia and ended up staying. The Baboon Island chimpanzee sanctuary that Janis directs is one of the largest ape sanctuaries in West Africa.

Janis's success in introducing chimpanzees to wild communities encouraged us to consider returning the baby to the Fongoli chimpanzee community, if possible. Janis emphasized the importance of keeping the baby from being in contact with people, due to the complications known to arise when infant chimpanzees grew attached to their human caretakers. These were the same measures outlined in a guide to ape reintroduction from the IUCN (International Union for the Conservation of Nature). We also wanted to safeguard the health of the wild population, if we were to reintroduce the baby to her group. We initially assumed that the hunters had killed the mother, which is common when an ape or monkey infant is captured; if this were the case, it would raise significant concerns about returning an unweaned infant to her group.

I flew from the United States to Senegal three days after I got Dondo's first call about the baby and arrived at his house in Senegal. It was arduous—I took a taxi straight from the airport after my overnight flight, then took a bush taxi for another fifteen hours or so, through the night, to reach Kédougou. By the time I arrived, the baby's eye was much better. I held her while she drank milk so that Dondo could administer drops to her eye, which I had brought with me from the US. I wore a surgical mask and sanitized my hands to reduce the likelihood that germs might be passed to her, as well as when contacting the baby's food (chimpanzees are very susceptible to many of the diseases humans are, given their close relatedness). Before I had left Dakar, I had bought a bag of small oranges on the roadside and, in addition to milk, I gave the baby small pieces of the oranges when I arrived and again the next morning. Even though she had never had anything resembling an orange, as far as I know, she ate it with gusto! After she had eaten, she lay down in the cage and gave what we call nest grunts, which the chimps give when they are settling down for the night. I slept on a bamboo bed on the other side of the hut, and I tried my best to nest grunt back to her.

The baby appeared to be in good health, and we had decided to take her out to Fongoli, which is only a 25-kilometer drive, to try to find her group, and see if her mother was still living. The next day, we drove out to Fongoli. We left for the field site with the baby in the cage around 5 a.m. and began a search around 6 a.m. of the area where chimpanzees had been seen a few days previously. Dondo and I split up to look for chimps; Michel stayed back at camp to watch and feed the baby. I found a group of chimpanzees at about 9 a.m. and saw that Tia was with them, but she was missing her infant. It wasn't until this moment that we realized the infant was Aimee. Baby chimps are hard to see—they cling to their mother's belly when they are young—and we didn't recognize the baby in the hunter's possession. When we encountered Tia, she was in a party of eight adult males and several adolescent males. She had gashes on her genital swelling where the hunters' dogs apparently had bitten her. I was in radio contact with Dondo, and he joined me and stayed with the group of chimpanzees while I jogged about a mile back to Fangoly village to get Aimee and Michel.

After Michel fed her one more time, I wrapped Aimee in a blanket and carried her to the chimpanzee group. I can still remember carrying her down the path from our camp at Fangoly village and passing Mbouly's wife, Sira. I had the blanket over Aimee's head, and I stayed back from Sira, but I turned so she could see I was carrying Aimee. Sira laughed—I don't imagine that's a sight you see very often—and I kept on going. Aimee had no difficulty clinging to my clothes, and I distinctly remembering her looking up into my face. I have no idea what she was thinking, and I was so nervous about the situation, I couldn't appreciate the fact that I was actually carrying a baby chimp.

As Michel and I approached the chimpanzees, we stopped and put Aimee in a burlap sack, hoping to place the sack on the ground some distance from the chimpanzee party and for them to find her. We did this because we were afraid the adult males especially might behave aggressively toward humans with an infant, as had happened when Janis reintroduced an infant to an unhabituated group of chimpanzees. Remarkably, none of the chimps ever reacted to us adversely, even though at least one of them saw us put Aimee down.

As we approached the chimps, we heard them hooting. I thought Aimee might respond, and we could place her on the ground then. However, she didn't vocalize, so we decided to move closer to the group, keeping tightly knit ourselves for our own protection. We moved to within about 10 meters of the chimps, who were feeding in a ficus tree in a woodland just above the Kerouani stream. The National Geographic Society had asked me to film the reintroduction, so I filmed while Dondo and Michel put the sack down and opened it so Aimee could see the chimpanzees. We saw some chimps observing this, and Mike—then an adolescent male who was at the edge of the feeding-tree crown—climbed down and approached Aimee. He came to within several meters of her, stopping, smelling, and looking intently at her before approaching to within a few feet of her. He stood on two feet, essentially holding his arms out and presenting his belly to Aimee for carrying her, and she climbed onto him, screaming in excitement.

Mike carried Aimee to the feeding tree, where Tia was just climbing down, probably after hearing Aimee cry. I saw Tia take Aimee from Mike, who had approached Tia. Other chimpanzees in the party quickly surrounded Tia and Aimee and appeared curious about her, smelling her intently. We could hear Aimee giving pant-grunt greetings. We couldn't see Tia or Aimee for several minutes in the mix of males, so I was a little tense. Surprisingly, none of the chimpanzees appeared alarmed at the fact that we had carried Aimee to them; there was no hair standing on end, warning vocalizations, or other behaviors to indicate stress directed toward us.

I stayed with the chimpanzee group for the rest of the day, from about 11 a.m. to 7 p.m., when the chimps made their night nests. None of the chimpanzees appeared to act differently toward me, although at least several of them had watched us place Aimee on the ground. I saw Aimee suckle and even play with her mother. When the chimpanzees traveled in the late afternoon, Mike carried Aimee to the nest site for Tia. When the chimps had started their long-distance travel at the end of the day, Tia couldn't keep up, and she frequently stopped moving to rest, checking her wounds and placing Aimee on the ground. One adult male, Diouf, lagged behind with her and stopped and waited when she rested. Finally, Mike returned, picked Aimee up from the ground, and carried her the rest of the way, just under a kilometer, to the nesting site, where he returned her to Tia. Mike's behavior demonstrated a clear example of empathy to me. (When I began writing a report about his behavior, I was surprised to learn how many scholars reserve feelings of empathy for our own species only; fortunately, this sentiment has changed significantly.)

The following day, Michel followed the chimpanzees while Dondo and I returned to Kédougou to meet with an official of the local wildlife department to discuss this incident. Michel saw Mike again carry Aimee for Tia during afternoon travel, this time for only about half a kilometer. It seems that Tia could keep up with the rest of the chimps in the morning while carrying her infant, but by the afternoon, she was again too tired to keep up. I then followed the chimpanzees from night nest to night nest, two days after we returned Aimee to her mom. Tia still moved slowly, stopping frequently during travel, but her wounds didn't bleed. Aimee appeared to be doing fine, and her eye looked almost completely normal.

Following the first several months after Aimee's return to Tia, the Fongoli team adjusted their data collection schedules to observe adult male focal subjects who traveled in the same party as Tia and Aimee to keep track of their progress. Approximately one month after the incident, Tia appeared to be fully recovered from her wounds, and she had resumed reproductive cycling. Tia and Aimee, as well as the parties in which they were found, were never seen to move back to the extreme southern part of their range. In fact, Tia and the parties in which she was seen remained near their core range area, not far from Fangoly village, rather than concentrating on an area called Potchokon with a permanent water source several kilometers east, as is usually the case during the late dry season. Tia and Aimee were seen frequently for the next three months (May through July) following their reunion, although observers reverted to the normal protocol that entailed following specific males according to our monthly schedule. During the six months following their reunion, neither Aimee nor Tia appeared ill.

One year later, however, it seemed that the chimps had not forgotten the incident. I was with a chimpanzee party that Tia and Aimee were a part of, and when the rest of the group moved south to travel on their usual route to the area where Aimee had been taken, Tia stayed behind, along with Tumbo—also a young mother at the time, with infant Cy. It would take years for all the chimps to begin using the southern area again. And recently, that area has become heavily disturbed by people from Kédougou town who are searching for gold. The town itself has expanded so much that Potchokon is now only about 3 kilometers from its outskirts—twice as close as it once was. The Fongoli chimpanzees understandably don't spend as much time at Potchokon anymore.

TOTO

In August 2012—more than three and a half years after we rescued Aimee—I got a call from Dondo that Tia had died and Aimee's little brother, two-month-old Toto, was in need of rescue.

When I got this call from Dondo, I was in my mother's hospital room at the MD Anderson Cancer Center in Houston, Texas.

I went into the bathroom in my mother's room and spoke with him about the rescue. I felt like I contemplated what to do for only seconds: I would ask my team in Fongoli to retrieve Toto. I have no doubt that if we had left Toto there at the end of the day to die, it could have alienated the people who live alongside the chimps (whom we inform about chimpanzee conservation), not to mention the men that work for me who consider the chimps to be like family. At that exact point in time, though, sitting in the hospital with my mom, the only reason I could think of was that I could not let someone else die. I could not let a baby chimpanzee die because he was supposed to die in the natural scheme of things, even if this is how scientific decisions should be made. At that moment in time, my heart could not bear the idea of letting this baby die. I didn't officially know that my mother's case was terminal until thirty minutes after the call from Dondo, when the doctor came and told us. However, I had already known deep down that my mother was dying, ever since we had heard the diagnosis of malignant melanoma some weeks before.

I spoke again with Janis Carter about how we might take care of Toto. We talked about factors like Toto's age and his health. Janis confirmed that the Baboon Island chimpanzee sanctuary was probably not an option for Toto, for at least two reasons—it was full to capacity, and Toto would need to be introduced with another chimp but there was no other chimp waiting to be introduced to the island. We also heard from Senegalese authorities that the zoo in Dakar would be a likely home for Toto, but we did not want to take that route. The zoo houses several chimpanzees but lacks the naturalistic habitat that sanctuaries or better-funded zoos in more developed countries have. Janis's efforts with reintroducing infants to other chimpanzee groups confirmed the likelihood that Toto could be adopted by wild chimpanzees if he were old enough to survive on his own, but he would need to be at least two years of age. We hoped to reintroduce him to his own group at Fongoli if possible, but we had to think of the welfare of those chimps first and foremost. If a reintroduced Toto were to approach unfamiliar humans because he lacked fear, he would put his entire group at risk.

It turned out that we took care of Toto for more than four years, until November 2016, when he was able to go to the Chimpanzee Conservation Centre, a sanctuary in Guinea. Toto lived at Janis Carter's Kédougou home, in his own room, under the care of two full-time caretakers, both named Ousmane. Fongoli Project personnel Dondo, Michel, Jacques Keita, and I also helped take care of Toto, which included trying to keep him busy and taking him out to the bush as often as possible so he wouldn't forget how to be a chimpanzee. Without chimpanzee role models, though, Toto went through some difficult phases. For a while, he moved around on two feet, and we were afraid he wouldn't go back to his knuckle-walking chimpanzee ways. But he did. For a very long time, he hated being out in the sun—something that savanna chimps would scoff at. He was better at learning to eat wild foods. In fact, we had to put him on a diet at one point because he got too fat. His large size worked against any hope of reintroducing him to his original chimpanzee group, though, as he looked more like a young male chimp than a baby male chimp, which could have drawn aggression from the adult males in the Fongoli group.

We had restricted the number of people around Toto, but he was still unafraid of humans—and anything else. There were too many people living in the Fongoli area now, because of the gold rush that had really escalated beginning around 2010, and it's likely that not all of them subscribed to the same cultural taboos against harming chimps that local people in Senegal did.

Unlike most orphaned chimps who have gone through a traumatic event upon capture, Toto went through a fairly uneventful rescue. I'm sure he was distressed as he lay with his dead mother that day, being comforted only by big sister Aimee when he cried. And he was dehydrated when the Fongoli team finally picked him up at around 5 p.m. and taught him how to drink from a bottle. But even though Toto was only a couple of months old, he had a relatively easy existence compared to almost all orphan apes who are rescued, where they are usually kept in less than ideal conditions, like small cages or in chains, after being stolen from their mother who has usually been killed in order to capture the infant. To this day, Toto is quite fearless, to the point that he doesn't always act like a young chimp should when he is introduced to other, bigger chimps at the sanctuary he now calls home.

With such an awful series of events, the ending of the Tia's story should be a happy one, and I am hopeful Toto's will bear that out since he is thriving in the Chimpanzee Conservation Centre sanctuary in Guinea. But Aimee disappeared about six months after her mother died. When the adult males in the group were aggressively competing over Lucille, who was in estrus, Aimee left the larger party of chimps, along with adult male Diouf and his sister, adult female Natasha and her kids, and some other females. The group of females and Diouf came back about a week later, but Aimee never did. We hoped she had merely been separated from the rest of the chimps and that she would show back up at some point, but this was wishful thinking.

Aimee was so plucky, but she was small for her age—too small, I think, to make it on her own. Even though I once saw her fight off much-larger adolescent female Sonja as Aimee tried to eat and was continuously harassed by Sonja, she was still vulnerable. After her mother died and she was alone, the males helped her out. Siberut and Bandit, the old guys, waited for her when she took longer to feed. Former alpha male Lupin, who may have been her father, also waited for her. Her human observers, Michel and I, would wait for her too. When she disappeared, I worried she fell behind with no one to wait for her. For a very long time, we all hoped she'd gotten lost and would turn up again, but I suspect she fell victim to a predator. It's easy to connect this death to the trauma of her youth and the death of her mother. Aimee was always somewhat fearful after being taken by hunters, never riding on her mother's back like older infants do, but remaining clung to Tia's belly. Her small size and her status as an orphan would leave her all the more vulnerable to dying of illness or being picked off by a predator.

For Tia's lineage, there is hope in her son Toto, who seems safe in the sanctuary. I think he will continue to thrive.

ETHNOPRIMATOLOGY

The way Fongoli chimpanzees react to humans, both directly and indirectly, is probably the most important thing to come out of our study, as we can use our knowledge to try and predict how chimpanzees in similar circumstances in this environment can and might deal with a growing human presence in this part of the world. Aimee's story is one example of what living in a human landscape might mean for a wild chimpanzee. Tia and Toto's story initially seemed less of a human problem. Tia had died from a lethal snakebite, and infant Toto would not have survived without her. But the circumstances of her demise and subsequent orphaning of infant Toto in a "human landscape" directly influenced my decisions about which actions to take to ensure Toto had the best future possible, either in the wild or in captivity. I think these two cases provide good examples of how the science we practice does not operate in a vacuum of human emotion, despite the fact that many scientists over the years have tried to operate this way, and primatologists like Jane Goodall have been criticized for allowing emotion to influence decisions, like inoculating chimpanzees against illness. I stand behind all the decisions I have made regarding the Fongoli chimpanzees, and I would make these same decisions again.

As far as chimpanzee study sites go, the Fongoli chimpanzee community holds a rare position: being one of the few that does not reside in an officially protected area, like a national park. Humans are a part of the chimps' lives and most likely have been for millennia. Even though the interactions between humans and chimpanzees are anthropologically of interest, we scientists have long tended to focus on primate groups that live in areas where humans have been excluded. In the early days of the Fongoli study, before the chimps were habituated to our presence, I spent more time with humans while trying to find the chimps in the first place, as well as establishing the project. The Fongoli Savanna Chimpanzee Project has come full circle after collecting years of observational data on the chimps, to where we have come to focus again on the intensified threats that chimpanzees face due to human impact.

Historically, most primatologists have conducted their research within national parks or other protected areas where humans had been excluded from living. In fact, this was thought to be the better place to conduct primatology studies especially if you were an anthropologist. Anthropology students were ultimately interested in evolutionary questions, and scholars thought that understanding nonhuman primates outside of areas where humans might have an influence would provide better insight into these primates' "natural" behaviors. Fortunately, we have changed our views. Given that primatologists have also begun to recognize the need to include anthropogenic impacts in research as a means to help conserve primate species, even anthropologists include "ethnoprimatological" studies in their research on nonhuman primates today.

Ethnoprimatology includes humans in the study of nonhuman primates and, in particular, considers the impacts of anthropogenic factors on primates' lives. Anthropogenic factors at Fongoli include horticulture and plant-food gathering, which have been historically common; the chimps seem to have adjusted well to such factors in southeastern Senegal. However, more-intensive agriculture and larger-scale artisanal mining, corporate-level mining, and intensive plant-food gathering for sale to markets in the capital city may ultimately impact the Fongoli chimpanzee community in such a way that threatens their existence.

In Senegal, most of the thousand or so chimpanzees live outside nationally protected areas. They are one of the few large animals left in the country living outside national parks, largely due to cultural taboos against eating them. Although Niokolo-Koba National Park, which is in the same region as Fongoli, is thought by many to be the premier game park in West Africa, a number of its species have declined in recent times. Elephants, the giant Derby eland, and lions are all found in relatively small numbers in the park, with fewer than ten elephants existing there (and thus in Senegal as a whole) today. But these and other species are either even rarer or completely absent outside this national park.

The critically endangered chimpanzee is one of the few species whose numbers were similar inside and outside park boundaries when we did our survey in the year 2000, which can be attributed to their protection via human cultural taboos in this country. When we first published the results of our survey of chimps within the Niokolo-Koba National Park and at some sites outside it, at least one reviewer was hesitant to approve our manuscript for publication because of our findings that chimpanzees outside protected areas in Senegal were doing just as well as those within the national park, at least in terms of their density. Although our science was sound—and ultimately published in that journal—that reviewer didn't want to highlight the notion that chimpanzees could live successfully alongside humans. Yet that is the case in Senegal. At least, it has been for a very long time.

Each of the cultural groups of people that live in the Kédougou region of Senegal has an origin narrative about how chimps came to be. Most of these detail the chimps somehow "falling from grace," as I put it. They began as humans, but then behaved badly—fishing on Friday, running away from a circumcision ceremony, living in caves—and were turned into chimpanzees. Chimps are therefore seen as too close to humans to eat, and this has been key to their conservation in Senegal. But, things are changing fast in this country, and even these taboos may not protect the chimps in the long run.

THE HUMAN LANDSCAPE AT FONGOLI

The human side of research on nonhumans can be more difficult than the day-to-day aspects of running a long-term research study site. The beginning phase of the Fongoli Savanna Chimpanzee Project involved a lot of human interaction in terms of cultivating relationships with the people who live alongside the chimps. Being a very introverted, shy person—and one who is really bad at learning languages, even worse at speaking them—involving humans in a big way in my research was very challenging for me. I calculated that I spent roughly one-third of my time looking for chimps during my first field season, one-third in transit be-

tween survey sites and sites where I met with people, and one-third inter-
acting with people who would help me, eventually work for my project,
or grant me permission to work in the area.

Dondo was the first person who worked on the Fongoli project, even
before it officially became the Fongoli Project, since he helped us out on
our initial surveys in the year 2000. So, when I returned to really kick
things off in 2001, I depended on him a lot. He served as translator between my very elementary French and the Malinke that our researcher/
guide Mbouly Camara spoke. When Dondo was growing up, he lived
part of the time in a Malinke village, where his sister lived. Mbouly grew
up in the Fongoli area; Dondo had also spent quite a bit of time in the
Fongoli area. But Dondo's home village is Thiobo (pronounced "chobo"
in English), which is also the home village of Jaques Keita and Michel
Sadiakho. We spent a lot of time visiting Dondo's family in Thiobo,
as well as the village of Djendji, where the local leadership lived. The
"chef" (chief) of Djendji village granted verbal permission for me to conduct my research in the area, so we visited Djendji, Thiobo and Fangoly
villages frequently.

These important but fairly relaxed meetings almost always involved
drinking tea. The black tea from China is brewed over coals in a little tea-
pot with a *lot* of sugar. Usually, there are three brews made from a little
box of tea about the size of a small bar of soap. The first round is very
strong; if I drink that first round of tea in the afternoon, I have a hard
time sleeping that night. This is the case even though it's served in little
glasses about the size of shot glasses. Rounds two and three are much
more mellow. I wish I would have kept track of how many glasses of tea
I drank that year.

I did usually keep track of how many meals we ate, which were also
fairly frequent: most days we would eat three lunches on our way home
or after returning back to camp from being in the field. During the initial years of the project, we were often staking out the water source near
Djendji, since it allowed us to sit back and watch the chimps descend
from the steep little hill that housed the natural spring at its base where
they drank almost daily during the late dry season. On the way home,
we would pass through Djendji village and eat with at least one family—
sometimes two. We would then eat again when we got back to Fangoly
village, at Mbouly's house. The hospitality was amazing, but Mbouly was

never convinced that I ate enough. Two very important words I quickly learned in the Mandingue language were those for "eat" and for "I'm full."

During short visits to my field site from the US, where I had a full-time teaching job, I might not spend much time in the field actually looking for chimps. My first time back to the project, in December 2001, I didn't even see chimps during my brief stay back in Senegal. I spent nine days in the field, but three of these were partial days and another two were spent in the bush at the Assirik site in Niokolo-Koba National Park. So, given we had only four full days in the field at the study site, it's not surprising we didn't see chimps. That's not to say we didn't collect some good data—we came across a group of fresh nests consisting of a party of eleven individuals. That same day, we found another fresh nest in the same general area. On our last day there, while Mbouly and I were doing a vegetation transect, Dondo, Mark Cook, and some other researchers found a fresh nest group of five to six individuals. Those were the days—finding fresh traces of chimps was really exciting! Actually seeing chimpanzees was an exceptional treat. At this point, it was very important to contact the chimps as much as possible, so they could begin to recognize us and realize we were harmless. But, since I made the decision early on to engage a small research team, to keep our footprint small—in terms of affecting the chimps as well as taking up resources from a relatively impoverished area, that puts a lot of pressure on team members.

My senior field assistant, Mbouly Camara, was instrumental to the success of the Fongoli Savanna Chimpanzee Project, and he remained a staunch supporter of my project even after he retired. He was the best field worker one could imagine. Upon arriving at Fangoly village one December, we found that Mbouly had a large, tumorlike swelling on the side of his neck that had been there for almost three weeks already. This didn't keep him from going out with us that first evening, and he still set a brisk pace. However, we took him with us to Kédougou that night and to the doctor the next day. The following day, we dropped him off at Sékoto, his home village, which lies along a main highway and is a few kilometers from Fangoly. He wanted to try and find some medical care there. The doctor in Kédougou had given him some pain pills, described the swelling as a tumor, and told him go to the regional hospital in Tambacounda. I wrote in my journal that I really was afraid Mbouly had some sort of a tumor that no amount of herbal practices would cure. I was wrong—

Figure 6.2. Aimee as a juvenile, with mother Tia. Photo by Joshua Marshack.

Mbouly went to a traditional healer who lanced the tumor with a white-hot metal instrument, and he eventually recovered.

Mbouly held such a vast amount of knowledge about the place he lived in. When I asked him about the number of chimps in the area, he said that long ago, there were many, many chimps. However, chimpanzees would raid people's beehives—the ones they weave from baskets—so people would shoot at the chimps. Then the "army" (I believe he was referring to the wildlife department) told the people they couldn't shoot at the chimps anymore, so now chimps are becoming numerous again.

One thing I noticed on our brief visit to Fangoly village that first year, in 2000, was that we were not served any wild game, or "bushmeat." Even though *bushmeat* is a common term that is almost always used to describe local people's practices, it implies poaching, an illegal and probably immoral means of obtaining meat. Whether the lack of wild game reflected a decrease in Mbouly's hunting because of his employment as a researcher or, perhaps, my US project manager's preferences for not eating meat, I am not sure. In fact, there is little wild game around Fangoly, which has led to changes in taboos on eating certain types of animals. For example, Dondo notes that while the current generation of Beudick people might include monkeys (but not chimps) in their diet, his parents' generation did not. This seems to me like a change in cultural taboos from the necessity of capitalizing on the wild game that was available. Monkeys are also major crop raiders in southeastern Senegal, while chimpanzees are not.

Although Mbouly's sons have worked for me in various capacities over the years, I was struck by the fact that they only had a small fraction of the knowledge their father had about the bush. I shouldn't have been too surprised, given the common shifts among generations to different lifestyles, but these men were also raised in a rural village, as their father was. I found myself more familiar with some plants, for example, than they were because their father had taught me about them. I regret that I didn't write down the different stories that Mbouly told his sons about the bush, which were translated to me in French from time to time.

Working in close collaboration with Senegalese people opened my eyes to the plight of many of the people living in this poor, rural area of the country. Infant mortality is high and not an uncommon event, for example. The southeastern part of Senegal was long seen as somewhat separated from the rest of the country, lagging in infrastructure and other resources. The region was relatively isolated from the capital of Dakar until a paved road was built in 1990. This road was a gift from the Saudi government, following a plane crash of Senegalese pilgrims on their way to Mecca—a gift that expressed the sorrow that the Saudi Arabian people felt in response to this tragedy. I spent much more time with humans than I ever anticipated before starting my research at Fongoli, and the dynamics between humans and chimpanzees in southeastern Senegal is perhaps more interesting than any other aspect of the Fongoli study. But it is largely overshadowed by the various new and different behaviors the chimps here exhibit.

CHIMPANZEES IN A HUMAN LANDSCAPE

Chimpanzees adjust to humans in different ways, from moving their yearly ranging paths to avoid gold mines, which Kelly Boyer Ontl showed in her doctoral research, to becoming habituated to the many motorcycles that now characterize the area. Fongoli apes react differently to different people, including those they must know to different degrees. For example, the chimps seem to know Omar, a farmer who works the land seasonally and lives in the town of Kédougou during the nonfarm-

Figure 6.3. Mbouly Camara (*far left*) and family with author Jill Pruetz (*second from right*) outside one of the huts at Fangoly village in 2014. Photo courtesy of the Fongoli Savanna Chimpanzee Project.

ing season, even though theirs is a complicated relationship. And Omar sometimes tries to involve us in his yearly struggle to keep them from eating all the fruit from a wild tamarind tree that grows near one of his fields.

Sometimes, I get to observe this relationship. Once, I was following a small group of chimps during the dry season. The chimps were lying in the shade under some *Saba* vines, and I was trying to find as much shade under nearby vines myself so I could still watch the chimps. We were all resting there when I heard the sound of running feet—running chimp feet. They would run for a number of steps, and then stop . . . and then start again. As they got nearer, I could hear a human some distance behind them, yelling in Puhlar. I don't speak Puhlar, so I'm not sure what he was saying, but I recognized Omar's voice. The next thing I know, the footsteps ended up at our *Saba* vines—they were chimpanzee Mamadou's, whose mother, Farafa, just happened to be one of the chimps in the small group I was observing. When Mamadou arrived, he stopped under the vines and let out a loud wraaa-bark (a threat) in Omar's direction. The other chimps joined Mamadou in wraaa-ing, and I heard the sound of Omar in the vicinity, apparently chastising the chimps in Puhlar. During a moment of silence, I said, "Ça va?"—a French greeting meaning "How are you?" that most people in that part of Senegal understand. After a beat or two, Omar calmly replied, "Ça va," as if he always ran into a woman hanging out with chimps under *Saba* vines, and then went on his way. I'm sure he couldn't see me or the other chimps, and I'm not sure why he was chasing Mamadou in the first place, but I was even more surprised that Omar could even halfway keep up with him, unless Mamadou wasn't too concerned about this particular human. At any rate, while the chimps are fearful of Omar and other farmers in the area, it's not the same as their fear of strangers like the herdsmen who started coming in 2006 with their big groups of sheep from the north of the country and from Mauritania, or of the periodic vehicles that come down the gravel roads within the chimpanzees' home range.

The chimps in Senegal face seemingly insurmountable obstacles associated with a booming human population, and conservationists have long recognized that they must appeal to local communities and stakeholders to help increase protection for other apes. Given the intelligence of chimps, they can exist alongside humans without the latter even knowing much about them. In November 2020, I received word from Dondo that someone had seen some chimps on the outskirts of Kédougou town at what used to be an isolated airstrip. This area is now part of the "suburbs" of Kédougou and is also where the police have set up a traffic checkpoint. A few chimps were seen there eating baobab, a main food at this time of year. Baobab trees are relatively scarce on the landscape in southeastern Senegal, so Fongoli chimps trapline these trees between November and January—moving between trees of the same species, sometimes covering long distances, up to 10 kilometers per day. (On average, chimpanzees at forested sites move about 2 kilometers per day, although some variation does occur, of course.)

My head researcher, Michel, went out the next day and identified elderly female Farafa, her infant, Arya, and adult male Mike. Michel hadn't seen Farafa for two months, and we always worry about elderly chimps when we haven't seen them for more than a month, so I was thrilled to hear that Farafa was still around. I wasn't thrilled to hear that she was so close to Kédougou, however. I've literally had nightmares where I dream the Fongoli chimps are doing things like crossing the main highway near

Kédougou to get to a baobab, hanging out at a ballpark, or nesting in my parents' backyard in Texas. Hearing that Farafa was hanging out close to the airstrip was a nightmare come true!

But Michel followed the trio for a few days and then lost them. We hoped they were heading back farther toward the interior of the Fongoli range and farther from Kédougou. Michel reported them back with the rest of the chimps in December. Farafa was in estrus, so it seems that she and Mike were in consortship. Chimps at other sites have been seen to engage in such consortships between adult males and estrus females, and they often go to the boundaries of their home ranges, so as to avoid the other adult males of the group. This can be dangerous if males from neighboring chimp communities discover them, given the nature of lethal attacks by community males on strangers. At a number of long-term chimpanzee study sites, adult males in particular patrol the boundaries of their home range and, if they encounter chimps from neighboring communities, and if they outnumber these "strange" chimps, they attack and can kill them. While chimpanzees at Fongoli do not engage in boundary patrols, in our case, hanging out so close to Kédougou town is equally dangerous as risking attack by hostile neighbor male chimps.

I often wonder if some of the males we have recorded as disappearing after intense dominance shifts are still around, at the boundaries of the Fongoli chimps' range. Farafa's son Mamadou disappeared after most of the adult males in the group attacked and killed Mamadou's close ally, former alpha male Foudouko (related in chapter 2). Mamadou was also chased by many of these younger males in the months following Foudouko's death, but I would see signs that he was still around for some time, such as finding a lone nest several hundred meters from where the rest of the community nested. I wonder if Farafa might have also been associating with Mamadou in an area like the one she was seen in near Kédougou town. Or perhaps Mamadou has since died. It is difficult to know the fate of chimps who have disappeared, although some scientists automatically assume such disappearances are deaths. We have discovered the fate of some of the Fongoli chimps that went missing from our records, and their stories are complicated—recall those I told about former alpha male Foudouko and about adult female Tia and her offspring.

· 7 ·
Conservation Threats and the Future of the Fongoli Chimpanzees

Whenever I arrive in Fongoli, I try to make a visit to Djendji village, the traditional seat of authority for the Fongoli Savanna Chimpanzee Project. I give my regards to the elders and other leaders there, and we share updates about our families, the village, and the chimp project. More than once during these visits, people have shared their preference for chimpanzees over baboons, another common primate found in the area. Baboons get a bad reputation because they are good at raiding people's crops. Chimpanzees raid crops in some areas too, but, in Senegal, the chimps usually ignore the common crops of peanuts, corn, and cotton. I've learned that the people in Djendji are happy to hear chimpanzees pant-hoot outside their village because it means there will be no baboons! For some reason, hearing this makes me feel like a proud parent. It's nice to know that people in the area associate chimpanzees with something positive. This is not always the case where chimpanzees live, and there are several human-related threats to chimpanzees in Senegal.

Since I started studying the Fongoli apes in 2001, the major threats to their livelihood have changed multiple times. When I initiated the project, I wondered if the existing human presence in the area—where chimps and humans had been living for perhaps millennia—could be adversely influencing the chimpanzees' health and well-being. In large part, this is because we primatologists and conservation biologists have long associated human presence with adverse effects on local wildlife. In recent years, however, we have begun to move away from this assumption to better understand the ethnoprimatological landscape, now recognized by primatologists as where human and nonhuman primates coexist and effect each other's livelihoods. Southeastern Senegal presents one such case where local human presence has contributed to the conservation of an endangered species, but with significant anthropogenic change in this region, there surely is a threshold that chimpanzees will not able to adjust to successfully.

As the human population increases and the livelihoods of the local population shift, so do the threats to Senegal's chimpanzees. At the start of my research, I wondered if the shifting agriculture negatively influenced these apes. It became apparent that a bigger stress was the intensive gathering of *Saba senegalensis* fruit by local people for sale to markets in the capitol city of Dakar. However, such a threat was not immediate—instead, it would reduce the propagation of this plant, which is important to both chimpanzee and human livelihoods in Senegal. However, by 2006, the appearance of sheep herds and herdsmen form northern Senegal and southern Mauritania produced an even bigger threat when they cut key tree species for browse for their herds. This, in turn, has been eclipsed by the gold rush that continues to boom in the region. In addition to artisanal and corporate-level mining activity, the associated increase in the local population has brought with it an increased pressure on natural resources. Woodcutters, for example, have depleted virtually all the larger kieno trees, which chimpanzees here feed on extensively and use as nesting trees, and which the galago prey they favor use as nesting sites.

CHIMPANZEE PROBLEMS ARE HUMAN PROBLEMS

The major issues faced by Fongoli chimpanzees are also major problems for people living alongside them. This is one of the reasons I established the nonprofit organization Neighbor Ape in 2008. For example, clean drinking water is what limits the chimpanzees the most in terms of where they can go and what they can eat during the dry season. Potable drinking water is also something rural people lack. This problem has become even more of a concern since the explosion of artisanal gold mining less than a decade ago in southeastern Senegal. Similarly, the woodland resources chimpanzees rely on are also crucial for rural people in this area of Senegal. As I mentioned above, one of the main foods the Fongoli chimpanzees depend on is a fruit from the *Saba senegalensis* vine. Local people, especially women, gather and sell this fruit during a time of the year when the previous year's crop yields are often depleted. In fact, it's a main source of cash money for women in rural villages, since men often control other resources that contribute cash to the economy.

Separating the research I do on the Fongoli chimpanzees from the development work that I help carry out with Neighbor Ape is impossible—without one the other wouldn't likely exist. I founded Neighbor Ape with the goal of giving back to the local community at Fongoli as well as working to conserve chimpanzees here. People in southeastern Senegal have been instrumental in preserving this large ape, so my mission statement acknowledges them: we "strive to conserve the chimpanzees in Senegal and to provide for the wellbeing of the people that live alongside them." As more primatologists and conservation biologists in recent decades have seen the need to involve local people in research and conservation, and disciplines like anthropology have recognized the value of conservation research, I find myself returning to the original type of study that I began at Fongoli—the interactions between people and chimpanzees on various levels—and wanting to tell the whole story behind this dynamic.

While people in this area of Senegal have protected chimpanzees via their long-held cultural taboos against hunting and eating them, our organization engaged almost thirty villages in discussions to encourage them to take additional steps to protect these species. Dondo Kante, our project manager, conducted all these workshops. After giving an initial presentation on basic chimpanzee behavior, he visited the village again on another day, meeting families and having tea with them and answering questions about chimpanzees. He came back a third time and gave a presentation on chimpanzee conservation, emphasizing the challenges that humans and chimps both experience in this part of the world. For example, we encourage people to recognize the importance of water and certain foods to chimpanzees and how they might reduce conflict over these resources. While hunting chimpanzees for food has not been done in recent history in Senegal, as far as I know, there are cases where people have taken the opportunity to capture an infant chimpanzee to be sold as a pet. As I've mentioned earlier, this almost always involves killing the mother, and we have taken special pains to educate people about how harmful this is, as well as how bad of a pet a chimpanzee makes.

Like elsewhere, young people in Senegal often spend less time in nature than their parents' generation. For this and other reasons, we strive to educate young people formally, especially girls, who have been traditionally left out of opportunities for formal education. Studies have shown that a higher education level for girls and women is linked with positive conservation efforts. But we encourage all young people to take an interest in nature and to recognize the special natural resources all around them in this area of Senegal. Most people here have no idea that the chimpanzees they share their land with are unique in many ways, so we emphasize chimpanzees as a national treasure for Senegalese people.

One of our biggest projects to date involved constructing a dormitory for village children in the regional capital of Kédougou, so that they may receive a better education. Other measures we have taken to support education and preserve the chimpanzee population include helping build and supply a pharmacy in Djendji, the largest village within the Fongoli chimpanzees' home range. Improving health for people in the area indirectly improves the chimpanzees' health as well. Not only does it help maintain a positive relationship with local people and the Fongoli Savanna Chimpanzee Project, it reduces the likelihood that chimpanzees contract illnesses from sick people. Since chimpanzees are our closest living relatives, they can get most of the same illnesses humans can, from the flu to COVID-19, but they often don't have the same immunity that humans typically do. During the initial years of the COVID-19 pandemic, we funded a local tailor in a mask-making project and distributed those masks to five different villages; through our grants from Neighbor Ape, we support the health costs of individuals who live in the Fongoli area. These projects illustrate the close partnership the Fongoli Savanna Chimpanzee Project has with the people of the region, who make my research possible.

In addition to forging relationships within local communities, I have advised corporate mining companies on how their actions can harm chimpanzees and have worked with them to educate Senegalese officials at various levels on the behavior of chimpanzees in their country. I have worked with organizations like Fauna & Flora International and the Biodiversity Consultancy to construct large-scale programs that benefit communities while preserving areas key to chimpanzees. For example, I worked with a mining company to identify key habitat areas for chimpanzees in the area, and these areas were then worked into a land management plan that the regional Senegalese community constructs, along with the aid of Fauna & Flora, to safeguard natural resources for future generations. In one case, as a result of understanding how important seasonal water sources are for chimpanzees, the company moved a road-building project to detour around a critical waterhole for a chimpanzee community that lives in the same region of southeastern Senegal as the Fongoli community. This same mining company funds large conservation organizations like Panthera and Fauna & Flora.

Not all the activities I have engaged in are immediately rewarding. After setting up a meeting between local authorities and the local wildlife authority in Senegal to talk about the increased tree cutting by migrating herdsmen from Mauritania in 2006, I learned that people in one of the villages mistakenly thought I prohibited middlemen from buying *Saba* fruit collected by residents to sell because trucks never arrived at the village that year. (We later heard that it was not profitable enough for the trucks to trade there that year.) While I had advised USAID officials that increased gathering of this wild fruit would ultimately prove unsustainable and would negatively affect the endangered chimpanzees in the country, I never advised a moratorium on local fruit gathering. Perhaps my recommendation played a part in the new restrictions that the wildlife authority placed on the gathering season for this fruit, restricting it in time. But, since I had students studying local human practices surrounding *Saba* as well as chimpanzee consumption, an inaccurate rumor began that I still encounter from time to time today. I don't know if that rumor hurt our project directly or not, but it has surely resulted in dissatisfaction on the part of some people who thought we had negatively affected their livelihoods, and I personally feel bad about the rumor. Our solution has been to increase communication, and thankfully, I think we've been able to counter this misinformation. At the time, however, the meeting I had with this particular village was one of the worst experiences I have had. I was extremely upset to hear anger from a village elder who assumed I had implied that the fruit-trade practice hurt the chimpanzees. I envisioned a broken relationship between my project and village leaders; fortunately, though, this did not come to pass. Ultimately, we seemed to overcome hard feelings on both sides. Fongoli Project manager Dondo Kante has always been instrumental in all communications.

Another example of a challenging conservation problem has to do with the popular idea that ecotourism is a universal solution to conserving an animal species and contributing to the local economy at the same time, and so providing incentive for animal conservation. This misconception stems from the assumption that the very successful mountain gorilla tourism projects in East Africa would produce the same results in Senegal with chimpanzee ecotourism. In Senegal, a number of issues prohibit what *seems* like an easy solution to conserving chimpanzees and developing economic sustainability for local people. In this country, most chimpanzees live alongside people. While the Senegalese have historically protected these apes via their cultural taboos against harming them, habituating chimpanzees to all humans leaves them vulnerable to people who don't hold the same values as people growing up in the area. It also causes chimpanzees to lose all their fear of humans, rather than just being okay with a few of them (in other words, our small research team and people associated with us) and still respecting humans more broadly and regarding them as dominant. The riskiest aspect of using chimpanzee ecotourism in rural areas of Senegal involves the risk for local people regarding attacks by chimpanzees. Unlike gorillas, chimpanzees eat meat—mainly, other primates. They do not distinguish between species and have attacked and killed human children in other parts of Africa, like Uganda, Tanzania, and Burundi. My biggest fear would be that an ecotourism project might result in chimpanzees who completely lose their fear of all humans, putting the humans they live alongside at great risk, especially children.

I have explained and elaborated on all the difficulties that chimpanzee ecotourism in Senegal faces, speaking to those from top-level USAID officials to mayors of local communities. I continue to try and educate people on the potential problems as well as alternative solutions to the conservation issues that surround Senegal's chimpanzees. Despite the struggles we encounter trying to conserve wild species, the rewards almost always make it worth the effort, even when things look grim.

WILDFIRES—A HUMAN PROBLEM?

I've mentioned throughout this book that people and chimpanzees in this region face many of the same problems. Yearly wildfires are one of the most pertinent examples. When we met with people from different villages in my ill-fated discussions about the impact of herdsmen on the Fongoli area, the number one threat to the environment and their livelihoods that people mentioned was wildfire. They also acknowledged that humans are usually the ones starting these fires.

Fongoli chimpanzees interact with wildfires yearly during the dry-season months, and while most of these fires are started by humans, the vegetation in the area is classified as fire adapted, indicating that wild-fires have characterized the landscape for some time. Moreover, the fires at Fongoli are extensive and regular—in fact, predictable for the chimps. They are much better than I am in predicting the movement of fires, and my number one rule when following chimps who encounter a wildfire is: "DO NOT LOSE THE CHIMPS!" They are expert at navigating around and even *through* fires—not very surprising, since fires are such a regular part of their environment. More than 70 percent of the Fongoli chimps' home range burns annually, largely due to human activity like clearing fields for planting, to clearing underbrush for better hunting, or clearing areas where artisanal gold mines are located and tall grasses make walk-ing and seeing the deep mine shafts difficult. The frequency and extent of wildfire at Fongoli is greater than at any other site where any apes are currently observed. But, unlike the disastrous and completely destruc-tive fires—for example, what orangutans in Borneo face when whole peat swamps burn for years, destroying their forested home—Fongoli chimps are able to live with wildfire challenges. This is largely because the veg-etation in their savanna home is fire adapted, such as trees having extra-thick bark that is impervious to wildfire. Yet, late-dry-season fires are also destructive, killing vegetation that humans and chimps alike depend on. And fires during the late dry season can be more dangerous because they burn rapidly through the exceptionally dry vegetation.

One of my most vivid memories includes a day out with the chimps when we encountered a late-dry-season wildfire that was larger than any I had seen, with flames climbing as high as a two-story building. I dis-tinctly remember my panicky burst through a group of chimps to get away from a wall of fire. They just sat and looked at me as if I had grown a second head, then got up and calmly walked away from the wall of flames. Although I think my interpretation of the threat was more severe than theirs, nevertheless, they did make an abrupt change from ascend-ing a towering baobab tree to escape the flames' reaches, to descending and moving ahead of it instead. It seemed the fire was too intense, with flames too high to risk sitting it out in a tall tree. In any case, I have yet to observe the chimps when they have been in a similarly threatening fire.

This was the same day I heard a very distinctive alarm call and an ex-aggerated display by the alpha male, Yopogon, "toward" the fire. Male chimpanzees display a lot, in fact. Displays are typically an attempt to intimidate someone—most often your group members, especially sub-ordinates or subordinates who are challenging your position. The alpha male's display was unusual only in that it was directed toward the fire.

There's a distinct "rain display" given at Fongoli and at other chimpanzee study sites during thunderstorms, consisting of almost slow-motion swaying and climbing through the trees, with charges at other groups members. At Fongoli, this is also one of the few times where humans are really included as targets, outside of the rare charging display by an alpha male or a rude adolescent male. This is exactly the type of display I witnessed as the wall of fire grew closer and closer to Yopogon. As the other chimps began filing out of the baobab tree, in apparent re-calculation of how high the fire's flames would reach, he remained in the tree, being the last to leave, and displaying in slow motion in the treetop, looking intently at the fire. In several other cases where I was with chimps when they encountered wildfires, they behaved similarly, watching large fires and appearing to calculate their movement accordingly. But these other fires were less severe, and the chimps either moved slowly around these fires or waited for the flames to pass before they moved through the still-smoking ash.

Since we began studying the chimps' behavior around fires, we also started learning how they use the burned areas of their range that year versus untouched areas. They tend to use the burned areas for feeding, although they don't take advantage of burned objects or insects, as monkeys in the area do. Rather, the burned areas at Fongoli seem to facilitate the chimpanzees' movements across their home range. Anyone here who has attempted to fight through the tall grass, which towers overhead during the months of September and October, can appreciate burned pathways.

Dr. Nicole Herzog led a study that included mapping the extent to which each new fire burned the chimps' territory and how they used these areas. After any new wildfire was detected, Fongoli researchers walked the edges of the fresh burn with a handheld GPS device, while a separate researcher followed an individual male chimpanzee subject's footsteps exactly as he moved across the landscape to measure his precise movements through or around these burns. When we compared the chimps' behavior in burned areas compared to untouched areas, we discovered that the chimps found fruit and fed faster in burned areas than in unburned areas. The extensive grassy understory of a savanna habitat makes it difficult to travel and find food for much of the year, especially during the early dry season before the fires burn through. It's not difficult to imagine early hominins attracted to wildfires for some of the same reasons, which may have led them to eventually learn to harness and even make fire themselves.

Additionally, Fongoli chimpanzees fed longer on honey in burned areas, but it took longer to detect beehives in these areas—perhaps because some fires destroy beehives, and in particular ones that are lower in tree trunks and branch cavities. Many of the beehives at Fongoli are found in the towering baobab trees, which are quite widespread here, and which house giant beehives in their big hollows.

FONGOLI'S FUTURE

Ironically, though people in southeastern Senegal have enabled chimpanzees here to continue existing alongside them even as other animals have been extirpated, the immigration of humans from other areas in West Africa may well lead to the demise of the Fongoli chimps and other communities of these critically endangered apes in Senegal. Beginning in 2006, we began to see some changes that led to an influx of additional anthropogenic, or human-related, disturbances at Fongoli that could potentially and negatively affect the chimps. The first of these was an influx of large herds of sheep, along with their shepherds, pushing down from northern Senegal and southern Mauritania, within the Sahelian zone just south of the Sahara. As the Sahara continues to creep southward and the vegetation in the Sahel region becomes increasingly sparse, pastoralists move farther to find forage for their sheep, goats, and cattle. The bigger change relates to the ongoing gold rush that took off in Senegal between 2006 and 2008 in the Fongoli region. These developments and related activity continue to threaten the long-term existence of the Fongoli chimpanzees and the rest of the chimp population in southeastern Senegal.

Although local people and chimps have been able to coexist in the Fongoli area for millennia, the chimps are now at a threshold of human disturbance that they may not be able to cross and survive. The future of the Fongoli chimps lies in the hands of their human neighbors. As primatologists, we have put our purely scientific research second to our efforts to join forces to try and conserve these apes, and we have also made connections with unlikely allies. For example, a key to the preservation of the savanna ecosystem is working with corporate-level gold-mining companies as well as local Senegalese stakeholders. The former have the resources necessary to make collaboration with the latter possible.

Again, the major issues faced by Fongoli chimpanzees are also major issues for people living alongside them. As I mentioned above, I advise corporate mining companies on how their actions can harm chimpanzees, and I work with them to educate Senegalese officials at various levels on the behavior of chimpanzees in their country. I have worked with organizations like African Wildlife Foundation and the Biodiversity Consultancy to construct large-scale programs that benefit communities but preserve areas key to chimpanzees. This has led to concrete results: for example, assistant director of the Fongoli Savanna Chimpanzee Project, Dr. Landing Badji, worked with teams to reseed areas that had been cleared for making roads and worked with the local communities to train eco-guards that identified illegal resource extraction in their areas. This same mining company works with large conservation organizations like Panthera and Fauna & Flora International. They work in particular to identify key habitat areas for chimpanzees in the area, as part of their offset program, and these areas are then worked into a land management plan that the regional Senegalese community constructs to safeguard natural resources for future generations.

In October 2023, I was invited to give one of the plenary talks at the Society for Francophone Primatologists, which held their annual conference in Dakar, Senegal. Preparing this talk really had me thinking deeply about the future of the Fongoli Project. My coauthors were Dr. Papa Ibnou Ndiaye of the University Cheikh Anta Diop in Dakar and Dr. Landing Badji of York University in Canada (and assistant director of the Fongoli Project). I titled the talk "Fongoli Savanna Chimpanzee Project: Past, Present and Future" and it represents somewhat of a turning point for me. While I had been thinking more and more about how to make the Fongoli Project sustainable since moving back to Texas, for some reason, speaking in public about it made it seem more real. It moved a private conversation with a few of us associated with the project to a public discussion of how to support and sustain it with Senegalese primatologists at the forefront.

Aside from the COVID-19 pandemic, the year 2023 was probably the year that I spent the least amount of time in the field at Fongoli. It really

Figure 7.1. Fongoli Savanna Chimpanzee Project researchers (*left to right*) Jacques Keita, Nazaire Bonang, and Jeremy Bindia at the 2023 Society of Francophone Primatologists meetings in Dakar, Senegal. Photo by the author.

did bother me—it has been very difficult to switch from planning my year around going to the field to planning my year around fundraising only. As scientists, we're always seeking funds for research, but in order to really secure the Fongoli Project, I believe that raising funds via the Neighbor Ape nonprofit organization is the way we will be able to establish an endowment for the project so that basic infrastructure costs, including salaries for Senegalese researchers, can be provided even in years when senior scientists don't secure grant funds.

I used to wonder how scientists could stop spending as much time in the field as they once did, but maybe I'm at that point in my career now too. In fact, I feel as if I should have been prioritizing this fundraising effort at least five years ago. But I hope I still have at least a decade to accomplish endowing the Fongoli Project. I never thought I would be spending so little time in the field—twenty years ago I was absolutely frantic to get to the field and spent, on average, about six months a year at Fongoli, for much of the time I worked at Iowa State University. I didn't have much of a life outside of work when I was in Iowa, either, but I was fine with that, and I have no regrets. What is surprising is that, since moving back to Texas, I actually started a better balance in my life. I've always loved animals—not just chimpanzees—so one of the first things I did when I moved back to Texas was to get two cows to mow my five acres (that was my excuse, anyway). My cousin, Larry Jacob, asked me if I wanted a Brahman calf whose mother had abandoned her. I drove Edith home in the backseat of my Prius (since traded in for a truck) and then had to get her a companion, so I bought Claire, a longhorn, the next day.

I have always loved horses, and I often wonder if I would have gone a different route in my life if I had owned one growing up. But I told myself long ago that I travel too much to care for horses. Then, my cousin Larry asked me if I wanted a free horse. I made the mistake of going to see the horse, a retired cowhorse named Smoke, and that was that. I had my first horse! I then bought a second horse from Larry because of course Smoke needed a companion because horses are herd animals, and the cows didn't really bond with Smoke, although he panicked when he couldn't find them in my pasture. Anyway, I've had a couple of cows, up to three horses and three dogs, and now a mule since I moved back to Texas in 2017.

I think the only way I was able to deal with not going to the field in Senegal for eighteen months during the pandemic was because I have my little ranch to keep me busy. But I never did think that I would spend as much time away from Fongoli as I have for the past few years. I understand better how and why people shift their focus as their research projects mature. I'm often surprisingly comfortable with it, if I can keep my eye on the prize, so to speak: endowing the Fongoli Project for future generations and still getting to spend time out in the field with the chimpanzees. But, after spending only two brief visits in Senegal in 2023, I am eager to figure out a way I can spend more time with the chimps again while still working hard on fundraising.

I envision the future of the Fongoli Savanna Chimpanzee Project as interrelated layers of people who have a stake in preserving the land and the animals living on it in southeastern Senegal. This includes the people who live alongside chimpanzees and share their landscapes, authorities who regulate resources used by both people and chimpanzees, Senegalese students and professionals devoted to conservation and science, and other people anywhere who care about the Fongoli chimpanzees. Key to this vision is the leadership of Dr. Papa Ibnou Ndiaye and Dr. Landing Badji. Their increasingly important roles in conservation and research at Fongoli will pave the way for a legacy of primatology and conservation in Senegal, which they have already begun to establish. The move to decolonize primatology and other scientific disciplines should not be a difficult task if scientists from the Global North can pay back the hospitality given to them by host countries in the form of resources available to them but more difficult for students and scientists from the Global South to acquire.

Senegal's motto of "Teranga" meant I was welcomed with open arms to study the fascinating chimpanzees of Fongoli simply because it was my dream to do so. I believe that the greatest legacy I could leave would be to return the favor and help young people from Senegal and around the world achieve their own dreams. If their dreams involve chimpanzees—well, that's even better.

Acknowledgments

The Fongoli Savanna Chimpanzee Project has been and continues to be a communal effort. Without the permission of the Department of Soils, Water, and Forests in Senegal, under the Ministry for the Environment and Sustainable Development, research on the Fongoli chimpanzees would not be possible. I am especially grateful for their genuine care and assistance in cases where we find deceased chimpanzees. The Fongoli team, sometimes along with local residents of the area and officials from the Kédougou Soils, Water, and Forests Department, treat the deaths of these critically endangered apes as they would those of persons they know. The chimps are buried for future scientific study of their bones, but in the same manner that people are buried in this part of Senegal, with as much reverence and solemnity. I have been absent for most of these events, but the ones I have attended have moved me immensely, both because of the loss of a chimpanzee I have loved and because of the respect paid to them by people who knew them or knew of them.

My own long and winding journey to begin research in Senegal was made possible by my family, friends, colleagues, and mentors. There are too many to list here, but I hope that almost everyone I have worked with over the years knows how much they have helped me fulfill my dream to study wild chimpanzees. If you have taken my classes, please know that you have helped me grow as an educator, a communicator, and a person. Graduate students have been and continue to be instrumental to the success of the Fongoli Savanna Chimpanzee Project. A short list of former students who have contributed to the Fongoli Project include Paco Bertolani, Stephanie Bogart, Kelly Boyer Ontl, Aaron Brownell, Clayton

Clement, Mark Cook, Maja Gaspersic, Susannah Johnson-Fulton, Stacy Lindshield, Alex Piel, Fiona Stewart, Peter Stirling, and Michel Waller, all of whom have played an important part in helping manage the project. In particular, Peter Stirling introduced me to Dondo Kante and the Fongoli site, so I owe him a great debt. If you have spent time out at the Fongoli field site or have helped work on the Fongoli chimpanzee database, please know you have contributed to the science and conservation of our project, and you are appreciated!

I have met many people throughout the years who are interested in conservation, and many of the folks I met on National Geographic Expeditions trips to see great apes have continued to keep in touch with me and to support the work we do in Fongoli. My friends and colleagues in Costa Rica are always in my thoughts as well and have contributed to my growth as a primatologist, and we are determined to link Senegalese students and conservationists with those in Costa Rica one day!

Film crews have included Fongoli chimpanzees in some exceptional documentaries, and I am immensely grateful that they have shown so many people around the world the lives of these amazing apes. I have a soft spot for the BBC *Dynasties* series in particular, given that it told the story of former alpha male chimpanzee David, one of my very favorite chimpanzees.

My family, on the Pruetz and on the Gallia sides, have always been supportive of and interested in what I do, and I am so very happy to be home in Texas again to reconnect with them and old friends I grew up with. It goes without saying that I couldn't do what I love without having the support of my family, and my parents and my brother have always been behind me. Tom LaDuke was also key in supporting me.

My friends have supported me unconditionally. I need to thank Susan Pavonetti in particular, as I would not be where I am without her initial and enduring friendship and support in so many ways of my work and of me as a person.

The people of southeastern Senegal in particular opened the door for me to follow my dreams of studying chimpanzees in a savanna landscape. It is an understatement to say that the Fongoli Savanna Chimpanzee Project would not have been possible without their assistance. They have truly made me feel like Fongoli is a second home to me, and they are among the most generous and caring people I've ever met. I have learned so much from the people in the Fongoli area—about wildlife, people, and myself.

Although I have mentioned members of the Fongoli Savanna Chimpanzee Project team throughout this book, I want to give my sincere thanks to Dondo Kante, Michel Sadiakho, Jacques Keita, Nazaire Bonang, Jeremy Bindia, Waly Camara, Fily Camara, Modi Camara, Sira Camara, Elhadj Camara, and their families for making us a part of their families and enabling the project to succeed. Dr. Landing Badji and Dr. Papa Ibnou Ndiaye have been and continue to be instrumental to the Fongoli Project.

Finally, I want to acknowledge anyone who enjoys learning about the Fongoli chimpanzees. I hope that I have been able to provide you an inkling of what it's like to be among these amazing creatures. They are truly worth knowing.

Further Reading

CHAPTER ONE

Gruen, L., A. Fultz, and J. Pruetz. "Ethical Issues in African Great Ape Field Studies." *ILAR Journal* 54, no. 1 (2013): 24–32.
Pruetz, J. D. "Studying Apes in a Human Landscape." In *Primate Ethnographies*, edited by K. Strier, 228–37. Pearson, 2013.
Stanford, C. B. *Chimpanzee and Red Colobus: Ecology of Predator and Prey*. Harvard University Press, 1998.

CHAPTER TWO

Achorn, A., S. Lindshield, P. I. Ndiaye, J. Winking, and J. D. Pruetz. "Reciprocity and Beyond: Explaining Meat transfers in Savanna-Dwelling Chimpanzees at Fongoli, Senegal." *American Journal of Biological Anthropology* 182, no. 2 (2023): 224–36.
Giuliano, C., F. A. Stewart, and A. K. Piel. "Chimpanzee (*Pan troglodytes schweinfurthii*) Grouping Patterns in an Open and Dry savanna Landscape, Issa Valley, Western Tanzania." *Journal of Human Evolution* 163 (2022): 103137.
Lehmann, J. and C. Boesch. "Bisexually Bonded Ranging in Chimpanzees (Pan troglodytes verus)." *Behavioral Ecology and Sociobiology* 57 (2005): 525–35.
Pruetz, J. D., and S. Lindshield. "Plant-Food and Tool Transfer among Savanna Chimpanzees at Fongoli, Senegal." *Primates* 53 (2012): 133–45.
Pruetz, J. D., K. B. Ontl, E. Cleaveland, S. Lindshield, J. Marshack, and E. G. Wessling. "Intragroup Lethal Aggression in West African Chimpanzees (*Pan troglodytes verus*): Inferred Killing of a Former Alpha Male at Fongoli, Senegal." *International Journal of Primatology* 38, no. 1 (2017): 31–57.
Sandel, A. A. "Male-Male Relationships in Chimpanzees and the Evolution of Human Pair Bonds." *Evolutionary Anthropology: Issues, News, and Reviews* 32, no. 4 (August 2023): 185–94.
Suzuki, A. "An Ecological Study of Chimpanzees in a Savanna Woodland." *Primates* 10 (1969): 103–48.

Wilson, M. L., C. Boesch, B. Fruth, T. Furuichi, I. C. Gilby, C. Hashimoto, C. L. Hobaiter, et al. "Lethal Aggression in *Pan* Is Better Explained by Adaptive Strategies Than Human Impacts." *Nature* 513, no. 7518 (2014): 414–17.

CHAPTER THREE

Boyer Ontl, K., and J. D. Pruetz. "Mothers Frequent Caves: Lactation Affects Chimpanzee (*Pan troglodytes verus*) Cave Use in Southeastern Senegal." *International Journal of Primatology* 41 (2020): 916–35.

Hunt, K. D. *Chimpanzee: Lessons from Our Sister Species*. Cambridge University Press, 2020.

Miller, McKensey. "Western Chimpanzee (*Pan troglodytes verus*) Use of Micro-climates in a Savanna-Woodland Environment: Behavioral Thermoregulation." Master's thesis, Texas State University, 2020.

Moore, J. "20-Savanna Chimpanzees, Referential Models and the Last Common Ancestor." *Great Ape Societies* 275 (1996).

Pruetz, J. D., and P. Bertolani. "Chimpanzee (*Pan troglodytes verus*) Behavioral Responses to Stresses Associated with Living in a Savanna-Mosaic Environment: Implications for Hominin Adaptations to Open Habitats." *PaleoAnthropology* (2009): 252–62.

CHAPTER FOUR

Ellison, G. "Behaviour and Ecology of the Northern Lesser Galago (*Galago senegalensis*)." PhD diss., Manchester Metropolitan University, 2022.

Gilby, I. C., Z. P. Machanda, R. C. O'Malley, C. M. Murray, E. V. Lonsdorf, K. Walker, D. C. Mjungu, et al. "Predation by Female Chimpanzees: Toward an Understanding of Sex Differences in Meat Acquisition in the Last Common Ancestor of *Pan* and *Homo*." *Journal of Human Evolution* 110 (2017): 82–94.

Pickering, T. R. *Rough and Tumble: Aggression, Hunting, and Human Evolution*. University of California Press, 2013.

Pruetz, J. D., and P. Bertolani. "Savanna Chimpanzees, *Pan troglodytes verus*, Hunt with Tools." *Current Biology* 17, no. 5 (2007): 412–17.

Pruetz, J. D., P. Bertolani, K. B. Ontl, S. Lindshield, M. Shelley, and W. G. Wessling. "New Evidence on the Tool-Assisted Hunting Exhibited by Chimpanzees (*Pan troglodytes verus*) in a Savannah Habitat at Fongoli, Sénégal." *Royal Society Open Science* 2 no. 4 (2015): 140507.

CHAPTER FIVE

Hart, D., and R. W. Sussman. *Man the Hunted: Primates, Predators, and Human Evolution*. Westview Press, 2008.

Lindshield, S., B. J. Danielson, J. M. Rothman, and J. D. Pruetz. "Feeding in Fear? How Adult Male Western Chimpanzees (*Pan troglodytes verus*) Adjust to Predation and Savanna Habitat Pressures." *American Journal of Physical Anthropology* 163, no. 3 (July 2017): 480–96.

Pruetz, J., T. C. LaDuke, and K. Dobson, K. "Savanna Chimpanzees (*Pan troglodytes verus*) in Senegal React to Deadly Snakes and Other Reptiles: Testing the Snake Detection Hypothesis." *bioRxiv*. Posted September 5, 2022. https://doi.org/10.1101/2022.09.04.506548.

CHAPTER SIX

Beck, B., K. Walkup, M. Rodrigues, S. Unwin, D. Travis, and T. Stoinski, eds. *Best Practice Guidelines for the Re-introduction of Great Apes*. World Conservation Union (IUCN) in collab. with the Center for Applied Biodiversity Science, 2007.

Ontl, Kelly Morgan Boyer. "Chimpanzees in the Island of Gold: Impacts of Artisanal Small-Scale Gold Mining on Chimpanzees (Pan troglodytes verus) in Fongoli, Senegal." PhD diss., Iowa State University, 2017.

Pruetz, J., and D. Kante. "Successful Return of a Wild Infant Chimpanzee (*Pan troglodytes verus*) to Its Natal Group after Capture by Poachers." *African Primates* 7, no. 1 (2010): 35–41.

Riley, E. P. *The Promise of Contemporary Primatology*. Routledge, 2019.

CHAPTER SEVEN

Herzog, N. M., J. D. Pruetz, and K. Hawkes. "Investigating Foundations for Hominin Fire Exploitation: Savanna-Dwelling Chimpanzees (*Pan troglodytes verus*) in Fire-Altered Landscapes." *Journal of Human Evolution* 167 (June 2022): 103193.

Lindshield, S., R. A. Hernandez-Aguilar, A. H. Korstjens, L. F. Marchant, V. Narat, P. I. Ndiaye, H. Ogawa, et al. "Chimpanzees (*Pan troglodytes*) in Savanna Landscapes." *Evolutionary Anthropology: Issues, News, and Reviews* 30, no. 6 (November/December 2021): 399–420.

Wessling, E., T. Humle, S. Heinicke, K. Hockings, D. Byler, and E. A. Williamson. *Regional Action Plan for the Conservation of Western Chimpanzees (Pan troglodytes verus), 2020–2030*. International Union for the Conservation of Nature, 2020.

Index

Page numbers in italics refer to figures.

Achorn, Angie, 28
adders, 92–93
African Wildlife Foundation, 126
aggression: affiliation and, 27; females and, 27, 40, 61, 69, 105; hunting and, 61–62, 67, 69, 73; intimidation and, 18; males and, 27, 32, 34, 37, 39–40, 42, 69, 73, 83, 85, 101, 105; Mamadou and, 18–19; Neighbor Ape and, 101, 105; risks and, 81–85, 88; sociality and, 27, 32, 34, 37, 39–42, 101, 105; unknown behaviors and, 7
Aimee: poachers and, 19, 85, 95; sociality and, *98*, 99–106, *111*
alarm barks, 36, 83–86, 88, 113
alarm call, 37, 83–84, 102, 123
alpha females, 31, 38, 60, 74–75, 89
alpha males, 20; conservation issues and, 123–24; fighting and, 19; heat issues and, 20, 46; hierarchy and, 19, 31, 35, 39; hunting and, 65, 68, 73–77; risks and, 85–86, 93; sociality and, 31, 34–42, 106, 115. *See also specific males*
American Sign Language, 99
American Society of Primatologists, 8, 11

anemia, 69
Ape Initiative, 11
Arya, 32, 40, 74, 114
Assirik, 4–5, 10, 12, 20, 110

Baboon Island, 99
baboons (*Papio papio*): bad reputation of, 117; conservation issues and, 117; heat issues and, 45, 52; hunting, 65–66; Neighbor Apes and, 99, 104; population density of, 20; risks and, 84; throwing stones at, 21
Badji, Landing, 126
bamboo, 5, 14, 55–56, 100
Bandit: heat issues and, 46, 50, 55–56; hunting and, 65–66; neighbor Aimee and, 105; risks and, 82–83, 90, 92; sociality and, 23, 37, 40, 105
Bantan community, 14, 18, 33–34, 37
Bantankiline Chimpanzee Project, 14
baobab trees: dry season and, *48*; food and, 5, 48, 86, 89, 114–15, 123–25; risks and, 86; sharing fruit of, 89; sociality and, 114–15, 123–25
bats, 52–53
BBC, 66

bees: allergies to, 88; chemical signals of, 89; dry season and, 88; fatal attacks of, 88; honey raids and, 8, 88–92, 125; lethality of, 88; risks and, 44, 81, 88–92

begging, 28, 64, 68–69, 74

behavior: aggression, 27 (*see also* aggression); begging, 28, 64, 68–69, 74; boundary patrolling, 32–34, 115; climate change and, 3, 7, 46; conservation issues and, 120, 124, 126; cooperative, 64; coordinated, 64; emotional, 19, 38, 62, 74–75, 102, 106; food sharing, 27–29; habituation, 4–11, 14, 19, 33, 38 (*see also* habituation); heat issues and, 45–47, 50–57; hunting, 1–2, 59–79; natural, 107; research methodologies and, 4–21; risks and, 85, 89–90, 93–94; rock throwing, 21, 83–85, 95; sociality and, 23–42, 97–115 (*see also* sociality); temper tantrums, 38, 74–75; theft, 28, 70, 73–74

Bertolani, Paco, 60

Beudick people, 13, 111

Bilbo: heat issues and, *48*; hunting and, 68; risks and, 84, 89, 91; social issues and, 31, 39–41

Bindia, Jeremy, *127*

Biodiversity Consultancy, 126

Blumenbach, Johan Friedrich, 51–52

Bo, *30*, 37–40, 73, 81

body temperature, 45–46

Boesch, Christopher, 25

Bonang, Nazaire, *127*

bonobos (*Pan paniscus*), 24, 27, 62–63

borescopes, 76–77

boundary patrolling, 32–34, 115

bush babies (*Galago senegalensis*): "chase and grab" style and, 65; hunting of, 1–3, 20, 54, 59–62, 65–79, 82; nesting of, 60, 65, 69, 71, 76–77, 100; risks and, 85

bushbucks, 21, 65–66, 84

bushmeat, 111

Camara, Jean Pierre, 13

Camara, Mbouly, *7*, *113*; as guide, 52;

109; harvest season and, 14; help from, 12–13; knowledge of, 111–12; lions and, 21; sociality and, 109–11; as village chief, 13

Camara, Sira, 101

Camara, Waly, 82

camera traps, 21, 48, 52–53, 83

Carter, Janis, 99, 104

caves: caches in, 21; heat issues and, 20, 47–48, 51–55; hunting and, 21; Kédougou narrative on, 108; risks and, 82–83; sociality and, 108

cheetahs, 72

Chimpanzee and Red Colobus (Stanford), 9

Chimpanzee Conservation Centre, 104–5

Chimpanzee Rehabilitation Project, 99

chimpanzees: African (*Pan troglodytes troglodytes*), 8, 27; as best-studied wild mammal, 7; East African (*P.t. schweinfurthii*), 26; endangered, 3, 45, 108, 118, 121, 125; habituation and, 1, 4–11, 14–19, 33, 38, 41, 51–54, 82, 88, 99, 101, 107, 112, 122; Nigerian (*P.t. vellerosus*), 27; Niokolo-Koba National Park and, 3–5, 10, 12, 20–21, 107–8, 110; sociality of, 23–42, 97–115; West African, 3, 7, 25–27, 47, 52, 64, 99, 107, 125

China, 52, 109

Clavette, Kerri, 10

climate issues: anthropogenic disturbances and, 3; behavior and, 3, 7, 46; dry season, 3 (*see also* dry season); ecosystem concerns, 7, 14, 21, 83, 126; heat, 3, 6, 12, 20, 23, 43–57, 81; humidity, 3, 12, 23, 44, 47–48, 93; rainy season, 15 (*see also* rainy season); risks and, 81

Cola cordifolia, 12

community, chimpanzee: Bantan, 14, 18, 33–34, 37; cohesiveness, 16, 24–29; conservation issues and, 119–21, 126, cost of rising in ranks, 32–34; East African, 25–29, 92, 122; Fongoli, 23–42, 46, 60, 87, 119, 121; food sharing and,

27-29; habituation and, 4-11, 14, 19, 33, 38; heat issues and, 46, 48; hunting and, 60-61, 70-73, 78; identifying individuals in, 15; Kanyawara, 4, 79; neighbor apes and, 99, 106-7, 115; risks and, 83, 87; sociality and, 23-42, 97-115

Congo, 8, 24

conservation issues: alpha males and, 123-24; baboons and, 117; behavior and, 120, 124, 126; community and, 119-21, 126; dry season and, 118, 123-25; ecosystem concerns and, 7, 14, 21, 83, 126; ecotourism and, 121-22; endangered species and, 3, 45, 108, 118, 121, 125; Fongoli and, 9-12; Fongoli Savanna Chimpanzee Project and, 117, 120-21, 126-29; food and, 118-19, 125; gold mines and, 118, 123, 125-26; habituation and, 122; historical perspective on, 12; infants and, 119; IUCN and, 99; mothers and, 119, 128; population growth and, 118; savannas and, 117, 120, 123-29; Senegal and, 117-22, 125-29; sociality and, 99, 103-5, 108, 114; Tanzania and, 122; thunderstorms and, 124; water and, 118-21; wildfires and, 81, 122-25; woodlands and, 118

Cook, Mark, 110

cooperation, 64

COVID-19 pandemic, 39, 120, 126

cows, 128

crocodiles, 48, 94

Cy, 29, 31, 40, 48, 49, 94-95, 103, *plate 6*

Dafoula, 87

Dakar, 13-14, 68, 100, 104, 112, 118, 126

David, *plate 4*; as alpha male, 20, 31-32, 35-42, 68, 73-77, 85-86; Farafa and, 18, 30-32, 36, 38, 40, 68, 73-75; Frito and, 18, 32, 38, 42; habituation and, 55-56; hunting and, 68, 73-77; risks and, 85-86; story of, 38-41

Dawson, 30, 86, 89, 92-94

dehydration, 42, 47, 52, 54, 81, 105

Derby elands, 107

Diouf, 30-31, 74, 84-85, 102, 105

Djendji, 5, 15, 21, 82-83, 109, 117, 120

DNA analysis, 18, 30

dogs, 128; domestic, 82-86; hunting and, 85, 97-98, 100; risks and, 82-83, 85-86; wild (*Lycaon pictus*), 21, 82

dry season, 113; conservation issues and, 118, 123-25; difficulty of, 2-3; Fangoly and, 103, 109; fighting and, 40, 42; food and, 25, 49, 54, 118, 125; heat index and, 3, 12, 20, 23, 47; hunting and, 59, 78; peak of, 15; risks and, 88; shade and, 12, 20, 43, 46, 113; springs and, 5-6, 90, 109; temperatures of, 3, 12, 20, 23, 43-44, 47, 50, 52, 55, 78; wildfires and, 123-25

ecosystems, 7, 14, 21, 83, 126

ecotourism, 121-22

elephant grass, 83

elephants, 107-8

Ellison, Grace, 71

emotion, 19, 38, 62, 74-75, 102, 106

empathy, 19, 102

ethnoprimatology, 106-8, 117

Eva, 2, 30

evolution: heat issues and, 45; hunting and, 61-63, 68; traditional research approaches to, 107

Fanta, 32, 51, 64, 68-69, 74-75

Farafa: curiosity of, 18; David and, 18, 30-32, 36, 38, 40, 68, 73-75; Fanta and, 32, 51, 64, 68-69, 74-75; heat issues and, 51-52; hunting and, 68, 70, 73-75; influence of, 29-31; Mamadou and, 18, 30, 32, 36, 82, 113, 115; play and, 23; risks and, 82-83; sociality and, 23, 28-31, 32, 36, 38, 40, 113-15

farmers: Fongoli and, 5, 10, 12, 14; Omar, 112-14; rainy season and, 14

Fatako stream, 5

Fauna & Flora International, 120-21, 126

females: aggression and, 27, 40, 61, 69, 105; alpha, 31, 38, 60, 74-75, 89;

females (*continued*)

drinking schedules and, 47; fighting and, 27, 30, 34, 105; habituation and, 16, 18–19, 30, 38, 41, 53; heat issues and, 50–53, 57; hierarchy and, 31, 39; hunting and, 1–3, 16, 26, 28, 51, 59–79, 82, 85; lactating, 47, 52; mating and, 31; mothers, 16 (*see also* mothers); nests and, 57, 60, 65, 69, 76–79, 115; play and, 9, 23, 29, 31, 36, 65–67; risks and, 82–86, 89, 93; sex ratios and, 25–26, 41; sociality and, 23–41, 99, 105, 114–15; tools and, 1, 2, 24–28, 59–61, 65–78, 82

fighting: dry season and, 40, 42; females and, 27, 30, 34, 105; males and, 19, 27, 30, 34; mobbing and, 83, 85, 90; mothers and, 30; throwing stones, 21, 83–85, 95

Fongoli: farmers and, 5, 10, 12, 14; finding sites in, 11–19; forests and, 1, 5–6, 10–13, 19; future of, 125–29; gold and, 3, 5, 17, 78, 103, 105, 112, 118, 123, 125; heat issues and, 6, 43–57; home range in, 26; human landscape at, 108–15; hunting in, 65–66; long-term study site for, 8–11; Niokolo-Koba National Park and, 4; research motivations for, 3–8, 15–21; savannas and, 1–7, 10–16, 19–20, 24–25, 29, 32–33, 36, 41–42, 45–48, 51, 63, 66, 81–85, 95, 104, 107–10, 113, 117, 120, 123–29; water and, 5–6, 12–20, 25, 40–52, 77–78, 82–87, 93–94, 118, 121; woodlands and, 5–6, 13

Fongoli Savanna Chimpanzee Project: Bertolani and, 60; Boyer Ontl and, 84; Camaras and, 82; conservation issues and, 117, 120–21, 126–29; Dondo and, 97, 98, 104, 109; founding of, 3–4, 14, 16–17, 19; risks and, 81–82, 84, 89, 95; Sadiakho and, 36, 104; sociality and, 107–10

"Fongoli Savanna Chimpanzee Project: Past, Present and Future" (Pruetz), 126

food: baobab trees and, 5, 48, 86, 89, 114–15, 123–25; begging for, 28, 64, 68–69, 74; competition over, 20; conservation issues and, 118–19, 125; dry season and, 25, 49, 54, 118, 125; field work and, 14; fruit, 5, 48–59, 77–79, 89, 113, 118, 121, 124; gathering, 10, 66–68, 107, 118, 121; heat issues and, 47, 49, 54, 56–57; honey, 8, 88–92, 125; hunting and, 77 (*see also* hunting); meat, 28–29, 61, 64–69, 73–75, 89, 111, 122; plants, 27–28, 77, 89, 107, 112, 118; risks and, 89, 99–100, 104, 107, 114; sharing of, 27–29; shea trees and, 5, 55; sociality and, 25–28, 32; termites, 8, 26, 28, 36–37, 56, 66–67, 70, 92; theft and, 28, 70, 73–74

foraging, 66, 83, 125

forests: Fongoli and, 1, 5–6, 10–13, 19; gallery, 5–6, 12, 48; heat issues and, 46, 48; hunting and, 60, 63–64, 77; marigot, 5–6; neighbor apes and, 114; risks and, 81, 83, 90; Sakoto ravine and, 6; sociality and, 25, 41, 114

Foudouko, *plate 1*; as alpha male, 19, 34–38, 42, 85, 114; death of, 36; risks and, 85; sociality and, 34–38, 41–42, 115; story of, 34–38

Frito, 1; David and, 18, 32, 38, 42; death of, 41–42; heat issues and, 47, 51; hunting and, 67–68, 71–72

fruit: food needs, 5, 48–59, 77–79, 89, 113, 118, 121, 124; *Nauclea*, 57; *Saba senegalensis*, 12, 49, 79, 97, 113–14, 118, 121

Gabon, 7–8

galagos. *See* bush babies (*Galago senegalensis*)

gallery forests, 5–6, 12, 48

Gambia River, 6, 48, 86–88, 94, 99

gathering: extractive foraging and, 66; horticultural, 107, 118, 121; hunting and, 10, 66–68

genetics, 10, 38, 62–63

genets, 52–53, 66, 72

Giringoto, 12

Giuliano, Camille, 25
gold: avoiding mines of, 112; conservation issues and, 118, 123, 125–26; Fongoli and, 3, 5, 17, 78, 103, 105, 112, 118, 123, 125; Kédougou and, 78, 103
Gombe Stream Research Centre, 10, 26, 29, 50, 68
Goodall, Jane, 10, 26, 106
gorillas (*Gorilla gorilla gorilla*), 8, 54, 122
grasslands, 5–6, 13, 46, 52, 84
Great Ape Trust, 11
green monkeys (*Chlorocebus sabaeus*), 83; caves and, 52; chimpanzees eating of, 34; as crop pests, 20–21; heat issues and, 52–53; hunting, 66
Guinea, 12–13, 104–5

habituation: chimpanzees and, 1, 4–11, 14–19, 33, 38, 41, 51–54, 82, 88, 99, 101, 107, 112, 122; conservation issues and, 122; females and, 16, 18–19, 30, 38, 41, 53; heat issues and, 51–54; limiting, 15–16; males and, 15–19, 30, 41, 53, 101; NSF grants and, 11; risks and, 82, 88; sociality and, 30, 33, 38, 41, 99–101, 107, 112
heat index, 3, 12, 20, 23, 47
heat issues: alpha males and, 20, 46; baboons and, 45, 52; Bandit and, 46, 50, 55–56; behavior and, 45–47, 50–57; Bilbo and, 48; caves and, 20, 47–48, 51–55; climate issues and, 81; community and, 46, 48; dehydration and, 42, 47, 52, 54, 81, 105; drinking schedules and, 47; dry season and, 43–57; evolution and, 45; Farafa and, 51–52; females and, 50–53, 57; forests and, 46, 48; green monkeys and, 52–53; habituation and, 51–54; homeothermy, 45; hormones, 46–47; hyenas and, 52; infants and, 51; Jumkin and, 46, 53, 56; leopards and, 52–53; Luthor and, 53, 56; males and, 46, 49, 50, 56–57; Mamadou and, 55; mothers and, 47, 51–52, 54; nocturnal activity and, 2, 54–57, 71; rainy season and, 47–51;

Sakoto ravine and, 48–53; savannas and, 6, 43–57; shade and, 12, 20, 43, 46, 113; Siberut and, 46, 50, 57; soaking pools and, 48–51; sociality and, 53, 57; thermoregulation, 45; thunderstorms and, 56; Tumbo and, 57; UV radiation, 45; water and, 43–52; wells and, 50–51; woodlands and, 42–56
Herzog, Nicole, 83, 124
hierarchy: alpha males, 19, 31, 35, 39; cost of rising in ranks, 32–34; females and, 31, 39
hippos, 48, 82, 86–88
homeothermy, 45
honey, 8, 88–92, 125
honey badgers, 52–53, 92
honeyguides (*Indicator indicator*), 92
hormones, 46–47
horses, 128
humidity, 3, 12, 23, 44, 47–48, 93
Hunt, Kevin, 51
hunting: aggression and, 61–62, 67, 69, 73; alpha males and, 65, 68, 73–77; baboons and, 65–66; Bandit and, 65–66; begging and, 28, 64, 68–69, 74; behavior and, 59–79; Bilbo and, 68; bush babies and, 1–3, 20, 54, 59–62, 65–79, 82; Camara and, 12, 82, 111; caves and, 21; "chase and grab" style, 65; clearing underbrush for, 123; community and, 60–61, 70–73, 78; cooperative, 1–3, 64; David and, 68, 73–77; dogs and, 85, 97–98, 100; dry season and, 59, 78; evolution and, 61–63, 68; Farafa and, 68, 70, 73–75; females and, 1–3, 16, 26, 28, 51, 59–79, 82, 85; food and, 47, 49, 54, 56–57; forests and, 60, 63–64, 77; gathering and, 10, 66–68; green monkeys and, 20, 66, 85; humans and, 10; hyenas and, 5; infants and, 16, 51, 59, 73–76, 82, 85, 97–100, 105–6, 119; Jumkin and, 71; leopards and, 21; lions and, 82; Luthor and, 70, 74; males and, 59, 65, 69–78; mothers and, 16, 51, 74–75, 97–100, 106, 119; poachers and, 14, 16, 19, 95,

hunting (*continued*)
111; rainy season and, 59, 71, 78; relational model and, 63; research motivations for, 61–64; Sakoto ravine and, 69; savannas and, 63, 66; seasonal patterns in, 78–79; Senegal and, 2–3, 10, 17, 20, 66, 78–79, 82, 85, 97, 119; Siberut and, 68, 75; small prey, 68–70; sociality and, 72; spear, 2, 59–61, 67–76, 82; taboos on, 10, 17, 111, 119; Tanzania and, 61, 65, 68, 71, 73; theft and, 28, 70, 73–74; tools and, 1–2, 17, 26, 28, 59–62, 65–78, 82–83; Tumbo and, 60, 68–71, 74–75; typical characteristics of, 64–65; warthogs and, 85, 97; woodlands and, 60
hydrophobia, 50
hyenas (*Crocuta crocuta*): heat issues and, 52; hunting and, 5; risks and, 81–85; strength of, 83; throwing stones at, 21

Indiana University, 51
infants: conservation issues and, 119; death of, 8; heat issues and, 51; hunting and, 16, 51, 59, 73–76, 82, 85, 97–100, 105–6, 119; mothers and, 16–19, 30, 35, 38, 51, 74, 94, 97–100, 103, 105–6, 119; neighbor apes and, 97–106, 112, 114; pet trade and, 16, 99; risks and, 82, 85–86, 94–95
Institut Pasteur de Dakar, 68
International Union for the Conservation of Nature (IUCN), 99
Ivory Coast, 25, 64, 72

jackals, 82, 85
Jacob, Larry, 128
Jumkin, *plate 5*; heat issues and, 46, 53, 56; hunting and, 71; risks and, 94; social issues and, 30, 37–40

Kante, Dondo "Johnny," 7; conservation issues and, 119, 121; as guide, 13; as manager, 97; sociality studies and, 97–104, 109–11, 114

Kanyawara community, 4, 79
Karamoko, 1, 67, 72
Keita, Jacques, 76–77, 109, 127
Kenya, 10, 92
Kerouani, 6, 52, 78, 101
Kharakhena, 52
Kibale National Park, 4–6, 73, 79
kieno (*Pterocarpus erinaceus*), 75–76, 118
K.L., 28, 39, 43–44, 56, 82, 85, 91

lactation, 47, 52
langurs, 52
Leakey Foundation, 11
lemurs, 2, 62
leopards (*Panthera pardus*): heat issues and, 52–53; hunting and, 21; Niokolo-Koba National Park and, 4; risks and, 81–83, 88; Sakoto ravine and, 21
Lex, 30, 31, 66, 86, 87, 90
Lily, 30, 34, 53, 85
Lindshield, Stacy, 27
lions (*Panthera leo*), 21, 82, 107
Loango National Park, 7–8
Lucille, 1, 31, 67, 72–74, 78, 105
Lucy, 99
Luna, 30, 74
Lupin, 30, 36, 65, 73, 91, 93, 106, *plate 3*
Luthor: heat issues and, 53, 56; hunting and, 70, 74; risks and, 84; social issues and, 30, 36–40

Mahale Mountains National Park, 2, 29, 61, 65, 72–73
males: aggression and, 27, 32, 34, 37, 39–40, 42, 69, 73, 83, 85, 101, 105; alpha, 19 (*see also* alpha males); boundary patrolling and, 32–34, 115; cost of rising in ranks, 32–34; fighting and, 19, 27, 30, 34; habituation and, 15–19, 30, 41, 53, 101; heat issues and, 46, 49, 50, 56–57; hunting and, 59, 65, 69–78; mama's boys and, 29–32; mating and, 31; nests and, 15, 37, 39, 55, 57, 65, 71, 84, 102, 115; play and, 23, 31, 36, 65, 85, 102; risks and, 84, 86, 93–94; sex ratios and, 25–26, 41; sociality and,

23–41, 97, 100–105, 115; tools and, 1, 26, 28, 59, 65–67, 70–78, 82
Mali, 52
Mamadou: aggression and, 18–19; DNA analysis of, 18; Farafa and, 18, 30, 32, 36, 82, 113, 115; heat issues and, 55; risks and, 82, 94; sociality and, 30–41, 113–15
mama's boys, 29–32
Mandingue language, 110
Maragoundi, 6, 13, 52, 55
Marchant, Linda, 10
marigot, 5–6
marmosets, 27
Matilda, 30, 74
mating, 31
Mauritania, 121
Max Planck Institute, 47
McGrew, Bill, 10
Miami University, 11
Mike, 36–39, 53, 101–2, 114–15
Miller, McKensey, 46
Miombo woodlands, 5
mobbing, 83, 85, 90
mongooses, 53, 65–66, 85
monitor lizards, 93
Moore, Jim, 52, 63
mosquitoes, 20, 56
mothers: conservation issues and, 119, 128; drinking schedules and, 47; fighting and, 30; heat issues and, 47, 51–52, 54; hunting and, 16, 51, 74–75, 97–100, 106, 119; infants and, 16–19, 30, 35, 38, 51, 74, 94, 97–100, 103, 105–6, 119; lactating, 47, 52; mama's boys and, 29–32; risks and, 94; sociality and, 27–40, 97–106, 111–13

National Geographic (journal), 17
National Geographic Society, 11, 101
National Science Foundation (NSF), 11
Nauclea fruit, 57
Ndiaye, Papa Ibnou, 14, 126
Neighbor Ape (organization), 13, 119, 128
nest grunts, 100

nests: construction of, 15; females and, 57, 60, 65, 69, 76–79, 115; males and, 15, 37, 39, 55, 57, 65, 71, 84, 102, 115
Niokolo-Koba National Park: Assirik and, 4–5, 10, 20, 110; research in, 3–5, 10, 12, 20–21, 107–8, 110

Omar, 112–14
Ontl, Kelly Boyer, 52, 84, 112
ophidiophobia, 94
orphans, 105–6
Oundoundu, 6

Panthera, 121, 126
pant-hoots, 20, 26, 37, 56, 117
pastoralists, 10, 125
patas monkeys (*Erythrocebus patas*), 10, 20, 66, 85–86, 92
Petit Oubadji ravine, 31
pet trade, 16, 99
Pickering, Travis, 61–62
plateaus, 5–6, 12, 23, 50, 82, 85, 90
play: females and, 9, 23, 29, 31, 36, 65–67; males and, 23, 31, 36, 65, 85, 102
poachers, 14, 16, 19, 85, 95, 111
Point D'eau, 15, 18, 90
porcupines, 52–53
predation: Aimee and, 106; diversity and, 4; human, 10; hunting and, 71 (*see also* hunting); mobbing and, 83, 85, 90; neighbor apes and, 106; pressure of, 20–21, 24, 71; prey dynamic and, 9; risks and, 81–92; rock throwing and, 21, 83–85, 95
Primate Conservation, Inc., 11
puff adders, 92–93
Puhlar language, 113–14
Purdue University, 27
pythons, 82, 93

rainy season: dry season and, 20 (*see also* dry season); farmers and, 14; heat issues and, 47–51; humidity and, 3, 12, 23, 47–48, 93; hunting and, 59, 71, 78; risks and, 93; seeps and, 15; sociality and, 23, 25–26; thunderstorms and,

rainy season (*continued*)
23, 56, 124; water and, 12, 15, 20, 25, 43, 47–51, 77–78
relational model, 63
risks: aggression and, 81–85, 88; alarm and, 37, 83–84, 102, 123; alpha males and, 85–86, 93; baboons and, 84; Bandit and, 82–83, 90, 92; bees and, 44, 81, 88–92; behavior and, 85, 89–90, 93–94; Bilbo and, 84, 89, 91; bush babies and, 85; caves and, 82–83; climate issues and, 81; community and, 83, 87; crocodiles and, 48, 94; David and, 85–86; dogs and, 82–83, 85–86; dry season and, 88; Farafa and, 82–83; females and, 82–86, 89, 93; Fongoli Savanna Chimpanzee Project and, 81–82, 84, 89, 95; food and, 89, 99–100, 104, 107, 114; forests and, 81, 83, 90; Foudouko and, 85; habituation and, 82, 88; hippos and, 48, 82, 86–88; hyenas and, 81–85; infants and, 82, 85–86, 94–95; jackals and, 82, 85; Jumkin and, 94; leopards and, 81–83, 88; lions and, 82; Luthor and, 84; males and, 84, 86, 93–94; Mamadou and, 82, 94; mobbing and, 83, 85, 90; monitor lizards and, 93; mothers and, 94; predation, 81–92; rainy season and, 93; rock throwing and, 21, 83–85, 95; Sadiakho and, 84, 86, 88, 92; Sakoto ravine and, 81, 83, 85, 93–94; savannas and, 81–95; Siberut and, 89; snakes and, 81, 88, 92–95, 106; sociality and, 81; tools and, 82–83; tortoises and, 82, 93; Tumbo and, 89, 94; turtles and, 82, 93–94; water and, 82–87, 90, 93–94; woodlands and, 84
rock throwing, 21, 83–85, 95
Ross, 17
Rough and Tumble (Pickering), 61–62

Saba senegalensis fruit, 12, 49, 79, 97, 113–14, 118, 121
Sadiakho, Michel, 56, *plates 7–8*; Fongoli Savanna Chimpanzee Project and, 36,

104; as head researcher, 33, 38–40, 71, 84, 86; risks and, 84, 86, 88, 92; snares and, 4; sociality and, 100–106, 109, 114–15
Sahara, 11, 125
Sakoto ravine: forests and, 6; heat issues and, 48–53; hunting and, 69; leopards and, 21; risks and, 81, 83, 85, 93–94; soaking pools and, 48–51; sociality and, 36–37
savannas: conservation issues and, 117, 120, 123–29; Fongoli and, 1–7, 10–16, 19–20, 24–25, 29, 32–33, 36, 41–42, 45–48, 51, 63, 66, 81–85, 95, 104, 107–10, 113, 117, 120, 123–29; grasslands, 5–6, 13, 46, 52, 84; heat issues and, 6, 43–57; hunting and, 63, 66; risks and, 81–95; Senegal and, 3, 10–12, 14, 20, 33, 46, 49, 66, 81, 83, 117, 126–29; types of, 5
Senegal: conservation issues and, 117–22, 125–29; Dakar, 13–14, 68, 100, 104, 112, 118, 126; future of, 125–29; heat issues and, 43–57; historical perspective on, 11–12; hunting and, 2–3, 10, 17, 20, 66, 78–79, 82, 85, 97, 119; languages of, 13; Mount Assirik, 4–5, 10, 12, 20, 110; Niokolo-Koba National Park and, 3–5, 10, 12, 20–21, 107–8, 110; rainy season and, 12, 20–21, 49; research in, 10–13, 17–18, 46, 81–82, 92, 99–100, 104–14, 118–19, 122, 127, 129; risks and, 81–95; savannas of, 3, 10–12, 14, 20, 33, 46, 49, 66, 81, 83, 117, 126–29; surveys in, 4, 10–12, 52, 108; "Teranga" as motto of, 129; woodlands and, 5–6, 13
sex ratios, 25–26, 41
shade, 12, 20, 43, 46, 113
sharing, 27–29
shea trees, 5, 55
Siberut: heat issues and, 46, 50, 57; hunting and, 68, 75; neighbor apes and, 105, risks and, 89; sociality and, 23, 37, 40, 105
snakes: ophidiophobia and, 94; puff

adders, 92–93; pythons, 82, 93; risks and, 81, 88, 92–95, 106

snares, 4

sociality: aggression and, 27, 32, 34, 37, 39–42, 101, 105; Aimee and, *98*, 99–106, *111*; alpha males and, 31, 34–42, 106, 115; Bandit and, 23, 37, 40, 105; begging for food, 28, 64, 68–69, 74; Bilbo and, 31, 39–41; boundary patrolling and, 32–34, 115; Camara and, 109–11; caves and, 108; conservation issues and, 99, 103–5, 108, 114; empathy and, 19, 102; Farafa and, 23, 28–31, 32, 36, 38, 40, 113–15; females and, 23–41, 99, 105, 114–15; Fongoli Savanna Chimpanzee Project and, 107–10; food and, 25–29, 32; forests and, 25, 41, 114; Foudouko and, 34–38, 41–42, 115; groups, 16, 18–21, 24, 26, 37–38, 41, 72, 81; habituation and, 30, 33, 38, 41, 99–101, 107, 112; heat issues and, 53, 57; human landscapes, 108–15; hunting and, 72; Jumkin and, *30*, 37–40; Luthor and, *30*, 36–40; males and, 24–42, 97, 100–105, 115; Mamadou and, 30–41, 113–15; mothers and, 27–40, 97–106, 111–13; pant-hoots, 20, 26, 37, 56, 117; play, 9, 23, 29, 31, 36, 65–67, 85, 102; rainy season and, 23, 25–26; risks and, 81; Sadiakho and, 100–106, 109, 114–15; Sakoto ravine and, 36–37; Siberut and, 23, 37, 40, 105; subgroups, 16, 25, 33, 40, 71, 84, 91; Tumbo and, *29*, 30–32, 36, 103

Society for Francophone Primatologists, 126

Sonja, *30*, 71–72, 105

spears: hunting and, 2, 59–61, 67–76, 82; kieno branches, 75–76, 118; tools and, 2, 59–61, 67, 69–70, 74–76, 82; typical, 75–78; wolo branches, 60, 76

spotted hyenas. *See* hyenas (*Crocuta crocuta*)

springs, 5–6, 90, 109

Spy in the Wild (BBC television program), 66

Stanford, Craig, 9

Stirling African Primate Project, 10

stone throwing, 21, 83–85, 95

subgroups, 16, 25, 33, 40, 71, 84, 91

subspecies, 3, 25–27, 63

taboos: changes in, 111–12; harming chimpanzees, 10, 17, 105, 108, 119, 122; hunting, 10, 17, 111, 119

Taï National Park, 25, 64, 72

tamarin monkeys (*Saguinus mystax*), 9, 27

Tanzania: conservation issues and, 122; Gombe Stream Research Centre, 10, 26, 29, 50, 68; hunting and, 61, 65, 68, 71, 73; Issa, 5, 25, 41; Mahale Mountains National Park, 2, 29, 61, 65, 72; Ugalla, 5; woodlands and, 41, 51

temper tantrums, 38, 74–75

"Teranga" motto, 129

termites: fishing for, 8, 26, 28, 36–37, 66–67, 70, 92; invasion of, 56; tools for getting, 8, 26, 28, 67, 70

territorial boundaries, 37–38

Teva, *35*, 99

Texas A&M University, 28

Texas State University, 8, 46

theft, 28, 70, 73–74

thermoregulation, 45

Thiobo, 13, 109

thunderstorms, 23, 56, 124

Tia, *30*, 85, 99–106, *111*, 115

tools: complex, 8; females and, 1, 2, 24–28, 59–61, 65–78, 82; hunting and, 1–2, 17, 26, 28, 59–62, 65–78, 82–83; kieno branches, 75–76, 118; males and, 1, 26, 28, 59, 65–67, 70–78, 82; risks and, 82–83; spears, 2, 59–61, 67, 69–70, 74–76, 82, 118; specialization of, 82–83; termite fishing, 8, 26, 28, 67, 70; tolerated theft of, 70; wolo branches, 60, 76

Toro Semliki Wildlife Reserve, 51

tortoises, 82, 93

Toto, *30*, 103–6

traplines, 91, 114

traps: camera, 21, 48, 52–53, 83; snares, 4; tree crowns and, 85

Tukantaba, 6, 84

Tumbo: heat issues and, 57; hunting and, 60, 68–71, 74–75; risks and, 89, 94; sociality and, 29, 30–32, 36, 103

turtles, 82, 93–94

Uganda, 4–6, 51, 73, 79, 122

University of California San Diego, 63

University of Cheikh Anta Diop, 14, 126

USAID, 121

US Fish and Wildlife Service, 11

UV radiation, 45

vervets, 10, 66

Vivienne, 32, 40, 82

vocalizations, 32, 102; barks, 36, 83–86, 88, 113; nest grunts, 100; pant-hoots, 20, 26, 37, 56, 117; screams, 20, 36, 56, 74, 77, 82, 86–90, 101

warthogs, 50, 56, 85, 97

water: conservation issues and, 118–21; dehydration and, 42, 47, 52, 54, 81, 105; drinking, 14–15, 40, 44, 47, 49–51, 82, 94, 105, 109, 118; drinking schedules and, 47; Fongoli and, 5–6, 12–20, 25, 40–52, 77–78, 82–87, 93–

94, 118, 121; heat issues and, 43–52; hippos and, 86–88; hydrophobia, 50; rainy season and, 15 (*see also* rainy season); risks and, 82–87, 90, 93–94; soaking pools and, 48–51; sources of, 5–6, 12, 15–17, 25, 40–47, 49, 52, 90, 103, 109, 121; springs and, 5–6, 90, 109; wells and, 50–51

water holes, 43–44, 82, 90, 121

wells, 50–51

Wenner-Gren Foundation for Anthropological Research, 11

Wessling, Erin, 47

wild dogs (*Lycaon pictus*), 21, 82

wildebeest, 24

wildfires, 81, 122–25

wolo (*Terminalia*), 60, 76

woodlands: conservation issues and, 118; Fongoli and, 5–6, 13; heat issues and, 42–56; hunting and, 60; Miombo, 5; neighbor apes and, 101; risks and, 84; Tanzania and, 51

Yopogon, 30, 123–24, *plate 2*

York University, 126

Zihlman, Adrienne, 42

Zoey, 30, 94

zoos, 62, 104